TL278 5 443 2004
0/3
Amas

Scie:
con

Springer
Tokyo
Berlin
Heidelberg
New York
Hong Kong
London
Milan
Paris

$119.95

Kakuro Amasaka

Science SQC,
New Quality Control Principle

The Quality Strategy of Toyota

With 186 Figures

250101

Springer

Kakuro Amasaka, PhD.
Professor
Aoyama Gakuin University
5-10-1 Fuchinobe,
Sagamihara-shi,
Kanagawa 229-8558, Japan

ISBN 4-431-20251-X Springer-Verlag Tokyo Berlin Heidelberg New York

Library of Congress Cataloging-in-Publication Data

Amasaka, Kakuro, 1947-
   Science SQC, new quality control principle: the quality strategy of Toyota / Kakuro Amasaka.
     p. cm.
   Includes bibliographical references and index.
   ISBN 4-431-20251-X (alk. paper)
   1. Automobiles--Design and construction--Quality control. 2. Automobile industry and
trade--Quality control. 3. Toyota Jidåsha Kabushiki Kaisha. I. Title.

   TL278.5.A43 2004
   629.2'34'0685--dc22

                           2004045314

Printed on acid-free paper

Springer-Verlag is a part of Springer Science+Business Media
Springeronline.com
Printing and binding: Hicom, Japan
SPIN: 10965252

Preface

As is represented by the world's top-level Japanese automotive manufacturers, Japanese manufacturers have been climbing to winning positions in global businesses. The driving force of their success includes responsiveness to diversifying market needs and elevating quality requirements, clarification of development concepts that facilitate new technology, application of technology to product planning and design, advanced production control systems that elaborately utilize state-of-the-art manufacturing technologies, and flexible, efficient production.

Many Japanese manufacturers succeeded in fulfilling these factors mainly because they utilize Japanese-style Lean Production Systems or the so-called Toyota Production System (TPS) and Japanese-style, scientific quality management approaches such as Statistical Quality Control (SQC) and Total Quality Management (TQM) in order to improve their corporate management technology.

As severe competition among manufacturers intensify both in Japan and overseas, action against quality problems that remarkably diminish customer satisfaction, and customer-first quality management, is becoming important. Scientific quality control methods that aim to optimize production processes for building in quality were undermined by the past expansion of quantity-oriented manufacturing through automation with massive and heavy equipment.

The author believes that the buoyant economy was the main cause of the prevalence of such manufacturing approaches. People became complacent about the economy and failed to continue SQC training and implementation, resulting in an overall decline in problem solving capability. This ultimately led to quality problem flow-out and unsolved, chronic in-process defects. Against this backdrop, there has been a growing need to develop a new SQC principle and its systematic application throughout organizations in order to enhance in-shop quality management through raising engineers' problem solving capabilities.

In principle, SQC is a methodology to help identify cause and effect correlations of apparently disorganized facts. SQC contributes to correcting work process and efficiently increasing work quality through correct fact analysis. In order to maximize the benefits of SQC, we should not only view SQC as an approach to statistically gain specific or partial conclusions, but also utilize it to gain general conclusions that contribute to technological development.

For this reason, the author established an SQC promotion cycle that continuously enhances existing technology. Furthermore, through an application of the cycle to engineers' business processes, the author also formulated a program that ensures elevation of technology and practical applications. The author proposes a new quality management methodology, Science SQC consisting of four core elements, as a demonstrative and scientific SQC methodology to enhance engineers' intellectual productivity.

The first core element is Scientific SQC that enables a scientific approach. The second is SQC Technical Methods that enable appropriate and speedy mountain climbing of problem solution. The third is TTIS (Total SQC Technical Intelligence System) that enables systematic SQC applications for inheriting and developing technology. The forth core element is Management SQC that enables identification of general technical conclusions through the clarification of cross-divisional business processes.

The author had been leading TQM promotion both in Toyota Motor Corporation and across Toyota group companies until 2000. During his career in Toyota Motor Corporation, the author had an opportunity to develop and implement next-generation TQM.

Although Japanese automotive manufacturers have become world-class business performers, in order for them to succeed in their global businesses (especially global production), they must continuously produce reliable products while responding to diversifying consumers' requirements and preferences.

For this reason, manufacturers must further strengthen their technical capability of product design and establish superior production engineering and production control systems in order to produce high-quality products at low cost. In this sense, quality management should be applied not only to downstream manufacturing plants but also to upstream design and development divisions. Managing and optimizing business process quality as well as product quality is becoming essential.

To overcome this challenge, manufacturers must evade conventional quality management approaches that largely depend on individual experience and intuition of key staff, i.e. engineers and management for enhancing technological capability. Instead, the author strongly believes in the necessity of establishing new quality management principles and core technology that comprehensively and effectively utilize the experience and know-how scattered across different levels, divisions and persons of the company.

Against this backdrop, the author implemented the new quality management principle, Science SQC so called Amasaka SQC Method as a core technology of Toyota's quality strategy in 1988. As a result, at each stage of the business process, from up to bottom in each division, the intellectual productivity of engineers and management was enhanced, contributing to production excellence in QCD (Quality, Cost and Delivery).

This book outlines the proposed Science SQC and verifies its validity through a number of demonstrative studies implemented in a leading company, Toyota Motor Corporation and Toyota group companies. Based on these successful applications of quality strategies, the author is promoting TQM-S (TQM by utilizing Science SQC) as a next-generation core TQM technology in Toyota Motor Corporation and Toyota group companies.

About the Author

Dr. Kakuro Amasaka was born in Aomori Prefecture, Japan, on May 5, 1947.

He received a Bachelor of Engineering degree from Hachinohe Technical College, Hachinohe, Japan, in 1968, a Master of Science degree from Pacific Western University, USA, in 1997, and a Doctor of Engineering degree specializing in Precision Mechanical and System Engineering, Statistics and Quality Control from Hiroshima University, Japan, in 1997.

Since joining Toyota Motor Corporation, Japan, in 1968, Dr. Amasaka worked as a quality control (TQM and SQC) consultant for many divisions. He was an engineer and manager of the Production Engineering Division, Quality Assurance Division, Overseas Engineering Division, Manufacturing Division and TQM Promotion Division (1968-1997), and the General Manager of the TQM Promotion Division (1998-2000).

Dr. Amasaka became a professor of the School of Science and Engineering, and the Graduate School of Science and Engineering at Aoyama Gakuin University, Tokyo, Japan in April 2000.

His specialties include: production engineering (Just in Time (JIT) and the Toyota Production System (TPS)), probability and statistics, multivariate statistical analysis, reliability engineering, and information processing engineering.

Recent research conducted includes: "Science SQC", New Quality Control Principle, "Science TQM" New Principle for Quality Management, "New JIT" New Management Technology Principle, "Customer Science" and "Kansei Engineering".

Positions in academic society and important posts: He is the author of a number of papers on strategic total quality management, as well as the convenor of JSQC (Journal of Japanese Society for Quality Control), JSPM (The Japan Society for Production Management), JIMA (Japan Industrial Management Association), ISCIE (The Institute of System Control and Information Engineers), JJSE (Japanese Journal of Sensory Evaluation), and other publications (e.g. ASQ and POM in USA). He has been serving as the vice chairman of JSPM (2003-), the director of JSQC (2001-2003), and the commissioner of the Deming Prize judging committee (2002-).

Patents and prizes: He acquired 72 patents concerned with quality control systems, production systems, and production engineering and measurement technology. He is a recipient of the Aichi Invention Encouragement Prize (1991), Nikkei Quality Control Literature Prizes (1992, 2000 and 2001), Quality Technological Prizes (1993 and 1999), SQC Prize (1976) and Kansei Engineering Society Publishing Prize (2002).

Address:
Office: 5-10-1, Fuchinobe, Sagamihara, Kanagawa-ken, 229-8558 Japan
Tel:+81 42 759 6313, Fax: +81 42 759 6556,
E-mail: kakuro_amasaka@ise.aoyama.ac.jp
Home: 3-3-19, Tsurumaki, Setagaya-ku, Tokyo, 154-0016 Japan
Tel and Fax: +81 3 3706 2095,
E-mail: amasaka@hn.catv.ne.jp

References (Quoted theses)

Chapter 1

K. Amasaka, (1998), Application of Classification and Related Methods to the SQC Renaissance in Toyota Motor, Data Science, Classification and Related Methods, pp. 684-695, Springer.

Chapter 2

K. Amasaka, (2000), A Demonstrative Study of a New SQC Concept and Procedure in the Manufacturing Industry: Establishment of a New Technical Method for Conducting Scientific SQC, International Journal of Mathematical & Computer Modelling, Vol.31, No. 10-12, pp. 1-10.

Chapter 3

K. Amasaka, (1999), A Study on "Science SQC" by Utilizing "Management SQC: A Demonstrative Study on a New SQC Concept and Procedure in the Manufacturing Industry, Journal of Production Economics, Vol. 60-61, pp. 591-598.

Chapter 4

K. Amasaka and S. Osaki, (1999), The Promotion of New Statistical Quality Control Internal Education in Toyota Motor: A Proposal of 'Science SQC' for Improving the Principle of Total Quality Management, European Journal of Engineering Education, Vol. 24, No. 3, pp. 259-276.

Chapter 5

K. Amasaka, (2003), Proposal and Implementation of the "Science SQC", Quality Control Principle', International Journal of Mathematical and Computer Modelling, Vol. 38, No. 11-13, pp.1125-1136.

Chapter 6
K. Amasaka, (1992), Future SQC which is useful for solution power Improvement of the Technological Subject, Collection of Activities Example using SQC Method to Improve Engineering Technologies, pp. 291-300, Japanese Standards Association, Nagoya QC Research Grpop.
K. Amasaka,(1993), SQC Development and Effects at Toyota, Quality, Journal of the Japanese Society for Quality Control, Vol. 23, No. 4, pp.47-58.

Chapter 7
K. Amasaka, (1999), A Proposal of the New SQC Internal Education for Management, Proceedings of the 15th International Conference on Production Research, Vol. 2, pp. 1147-1150, Limerick, Ireland.

Chapter 8
K. Amasaka, (1995), A Construction of SQC Intelligence System for Quick Registration and Retrieval Library: A Visualized SQC Report for Technical Wealth, Stochastic Modelling in Innovative Manufacturing, Lecture Notes in Economics and Mathematical Systems, Vol.445, pp. 318-336, Springer.

Chapter 9
K. Amasaka, A. Nagaya and W. Shibata, (1999), Studies on Design SQC with the Application of Science SQC: Improving of Business Process Method for Automotive Profile Design, Japanese Journal of Sensory Evaluations, Vol. 3, No. 1, pp. 21-29.

Chapter 10
T. Takaoka and K. Amasaka, (1991), Analysis of Factors to Improve Fuel Efficiency, (In Japanese) Quality, Journal of the Japanese Society for Quality Control, Vol. 21, No. 1, pp. 64-69.

Chapter 11
K. Amasaka and S. Osaki, (2002), Reliability of Oil Seal for Transaxlc: A Science SQC Approach in Toyota, Edited by Wallace R. Blischke & D.N.Prabhakar Murthy, Case Studies in Reliability and Maintenance, pp. 571-581. John Wiley & Sons, Inc.,

Chapter12

K. Amasaka, H. Nakaya, K. Oda, T. Oohashi and S. Osaki, (1996), A Study of Estimating Vehicle Aerodynamics of Lift: Combining the Usage of Neural Network and Multivariate Analysis, (in Japanese) The Institute of Systems Control and Information Engineers, Vol. 9, No. 5, pp. 229-235.

Chapter 13

K. Kusune, Y. Suzuki, S. Nishimura and K. Amasaka, (1992), The Statistical Analysis of the Spring-back for Stamping Parts with Longitudinal Curvature, (in Japanese) Quality, Journal of the Japanese Society of Quality Control, Vol. 22, No. 4, pp. 24-30.

Chapter 14

K. Amasaka, Y. Mitani and H. Tsukamoto, (1993), 'Study of Quality Assurance to Protect Plating Parts from Corrosion by SQC: Improvement of Grinding Roughness for Rod Piston by Centerless Grinding, (in Japanese), Quality, Journal of the Japanese Society for Quality Control, Vol. 23, No. 2, pp. 90-98.

Chapter 15

K. Amasaka and H. Sakai,(1996), Improving the Reliability of Body Assembly Line Equipment, International Journal of Reliability, Quality and Safety Engineering, Vol. 3, No. 1, pp. 11-24.

Chapter 16

K. Amasaka and H. Sakai, (1998), Availability and Reliability Information Administration System "ARIM-BL" by Methodology in "Inline-Online SQC", International Journal of Reliability, Quality and Safety Engineering, Vol. 5, No. 1. pp. 55-63.

Chapter 17

K. Amasaka and H. Sakai, H (2002), A Study on TPS-QAS When Utilizing Inline-Online SQC: Key to New JIT at Toyota, Proceedings of the Production and Operations Management Society, San Francisco, California, CD-ROM, pp. 1-8.

Acknowledgements

These days, there have been cases where leading companies undergo hard times with unanticipated quality problems, and cases where companies loses sight of customer needs and fail to catch up with technological advancements, placing their corporate existences in danger. In such circumstances, the author believes that management should aim for justifiable customer first quality management. The key to achieving customer-oriented quality management is establishment and utilization of demonstrative scientific methodology.

The author continues to propose a new quality control principle, Science SQC, as a next-generation quality management principle, and verified its validity through a number of demonstrative studies. During his time working at Toyota Motor Corporation (1968-2000), the author served as the general manager of the TQM Promotion Division from 1988, and lead Toyota's quality management mainly through effective applications of Science SQC to Toyota group companies. The renowned "LEXUS" brand represents the success of the group-wide Science SQC application that contributed to production of quality excellence in vehicles.

This book is a compilation of past studies and referees' comments released in major academic journals, and explains the comprehensive theory and practices of Science SQC so called Amasaka SQC Method that have been widely used and are evolving as Toyota's new scientific quality management methodology.

The author would like to acknowledge the generous support received from the following researchers. All those at Toyota Motor Corporation that assisted with the author's research, especially, Mr. H. Sakai, Mr. A. Nagaya, Mr. T. Takaoka, Mr. H. Nakaya, Mr. K. Kusune, and Mr. Y. Mitani. Dr. S. Osaki of Nanzan University for the many comments and suggestions offered. Mr. T. Hayatsu of Cardinal System Corp. and Springer-Verlag Tokyo Inc. for their continued support and corporation in publishing this book.

Contributing Authors

Akihiro Nagaya	Toyota Motor Corporation, Toyota, Japan
Hirohisa Sakai	Toyota Motor Corporation, Toyota, Japan
Hiroyuki Nakaya	Toyota Motor Corporation, Toyota, Japan
Hitori Tsukamoto	Toyota Motor Corporation, Toyota, Japan
Kouji Kusune	Toyota Motor Corporation, Toyota, Japan
Kazunori Oda	Toyota Motor Corporation, Toyota, Japan
Shingo Nishimura	Toyota Motor Corporation, Toyota, Japan
Shunji Osaki	Nanzan University, Nagoya, Japan
Tetsuya Oohashi	Toyota Motor Corporation, Toyota, Japan
Toshifumi Takaoka	Toyota Motor Corporation, Toyota, Japan
Wako Matsubara	Toyota Motor Corporation, Toyota, Japan
Yuushi Mitani	Toyota Motor Corporation, Toyota, Japan

Contents

4 Technological Quality Strategy in Toyota

4-1. Intellectual Product Development

4-2· Intellectual Production Technology

Structure

This book provides overall explanations of the new quality control principle, Science SQC so called Amasaka SQC Method, and reports how quality strategies based on Science SQC were implemented at Toyota Motor Corporation. This book mainly consists of quotations from the author's theses announced and publicized in Japan and overseas and has the following structure.

Structure

1, Introduction, explains the new quality control principle, Science SQC and outlines Toyota's quality strategies.

2, Science SQC, New Quality Control Principle, includes Chapter 1, SQC Renaissance as an application of the SQC Promotion Cycle; Chapter 2, Scientific SQC approach, as an application of the new SQC Technical Methods; Chapter 3, "Science SQC" Proposal, as an effective implementation of Management SQC; Chapter 4, Science SQC Implementation, as stratified SQC training to enhance engineers' problem solving capability; Chapter 5, Science SQC, A New Quality Control Principle, as a proposal for next-generation TQM, TQM-S, which played a key role in Toyota's quality strategy.

3 outline Toyota's quality strategies that utilized Science SQC. 3, entitled Human Resource Development and Practical Outcomes of Science SQC in Toyota includes Chapter 6, SQC Promotion Cycle Activities, for engineers' human resources development and practical results; Chapter 7, Management SQC, lead by general managers and managers; and Chapter 8, TSIS-QR system, a sub-core element of SQC integration network.

4 entitled Technological Quality Strategy in Toyota introduces specific studies. 4-1 highlights intellectual product development through the introduction of studies in product plan design, development design, reliability design, and CAE design from Chapters 9 to 12. 4-2 highlights intellectual production technology through introduction of studies in production engineering, process design, production preparation, production process, and process control from Chapters 13 to 17. Finally, 5 provides overall conclusion of the topics covered in this book.

1. Introduction

1. Introduction

1.1 The Proposal of "Science SQC", New Quality Control Principle

The development of the manufacturing industry in Japan is attributable to manufacturers who developed Japanese quality control method by learning Statistical Quality Control (SQC), which represent the scientific quality control methodologies advocated by Dr. Deming [1], Dr. Shewhart [2] et al. They incorporated these methodologies into their activities for improving quality engineering through Total Quality Management (TQM) by Dr. Juran [3], et al.

An overview of the latest state of Japanese quality management indicates an expansion of quantity-oriented manufacturing through automation with massive and heavy equipment. Such a trend has consequently diminished the Scientific Quality Control Method by which managers, supervisors, engineers and workers collaboratively build quality on the manufacturing line by optimally managing 4M-E (Man, Machine, Material, Method and Environment) [4-5].

Even among leading enterprises, it is apparent that engineering staff and managers are turning away from quality control training and education. SQC, which forms the core of quality management, is less frequently applied while staff and managers' problem solving skills have been weakening. As a result, it is often observed that companies fail to identify problems and take prompt measures for quality problems [6-7]. Thus, market quality problems and defects in the process remain unsolved, becoming chronic without a sign of decrease. Some manufacturers are observed to repeat trial and error in their frantic effort for solutions. Most of those chronic engineering problems are caused by poor job performance of associated sectors, from the upper to lower stream in the organization.

Under the circumstances, to reinforce quality control in the production lines of Japan, an application of new SQC from a management point of view is fervently desired along with the reconstruction of a systematic, organized application, particularly to contribute to enhancing engineers' problem solving capabilities [8]. On the basis of the above-stated points, changes in the environment surrounding Japanese enterprises, mainly in manufacturing, once again demands corporate efforts to enhance engineers' intellectual productivity to further increase their engineering development capabilities for best application to merchandise development.

To create merchandises which meets customers' satisfaction, it is important to work under optimal conditions that enhance engineers' job quality while eliminating any problems. To do this, the effectiveness of applying SQC, which is a core technique of TQM, must be recognized again. SQC must be utilized from a scientific viewpoint in order to contribute to the evolution of technology for solving not only explicit engineering problems but also foreseen, implicit technological tasks.

In principle, SQC is a methodology to help identify cause and effect correlations of apparently disorganized facts. SQC contributes to correcting work process and efficiently increasing work quality through correct fact analysis. In order to maximize the benefits of SQC, we should view SQC not only as an approach to statistically gain specific or partial solutions, but also utilize it to gain general solutions that contribute to technological development.

In order to go beyond reactive problem solution and proactively tackle engineering issues by seeking "general solutions", continuous applications of SQC and the obtained general solutions are essential. For this, implementation of so-called "SQC Renaissance" [9] is necessary. SQC Renaissance is an initiative to ensure revival of SQC and promotion of new SQC methodology organization-wide through establishment of systematic study scheme and promotion cycle. SQC Renaissance ultimately aims that SQC is regularly incorporated in engineering business. To concretely develop this, it is important to create a program for effectively operating the SQC Promotion Cycle (Implementation-Practical Effort-Education- Growing Human Resources) to promote a systematic and organized operation of SQC, and a system and practice to allow the cycle to spiral upward.

Furthermore, the author [10-12] established a positive scientific SQC application to contribute to improving technology. The author proposed a new 4-core quality control principle, "Science SQC", New Quality Control Principle as a new systematic method of applying SQC to contribute to the improvement of management techniques by enhancing the intellectual productivity of engineers and managers.

The first core refers to Scientific SQC. It enables a scientific approach during every stage of the action process, from problem structuring to target achievement, by clearly identifying an ultimate goal. The second core is the SQC Technical Methods, which enables correct and prompt mountain climbing of problem solution. The third core is TTIS (Total SQC Technical Intelligence System) for systematic SQC application, which contributes to building technical assets and helps pass them along for development. The fourth core, Management SQC, is a methodology for turning the implicit knowledge of intra-division business process into explicit knowledge through scientifically examining the gap between theory and reality that exist in an organization. This is based on a concept that the analysis of technical problems (scientific analysis of discrepancies in the principles and/or rules) is reduced to the problem of organization.

By developing and promoting these 4 cores, the proposed Science SQC so called Amasaka SQC Method can contribute to creating products that excel in QCD (Quality, Cost and Delivery) by improving engineers' job quality in each stage of divisional business process, from the upper stream to lower stream. This has been verified by Toyota, an advanced enterprise, and its group companies [13-15].

1.2 The Quality Strategy of Toyota

In the opinion of most countries, the Japanese manufacturing industry seems to have high productivity and good product quality. Particularly, the automotive industry-a general assembly industry supported by a number of related industries and associated technologies- has advanced to the top level in the world.

This is attributable to two things: the efforts made toward building vehicles of good total balance and the realization of high quality assurance with priority given to customers through collaborative creation activities of the vehicle and parts manufacturers [16-17]. The major factors for success thus far are that (1) in engineering development, new technologies were developed under clearly defined development concepts, which were reflected in product planning and design, and (2) in manufacturing, a flexible, efficient production system was realized by developing manufacturing lines to a high-level production management with full application of advanced production and processing technologies.

What is significant here is that the high quality production is based on a customer first concept. It was promoted by integrating a Japanese production system as represented by the so called hard technology of the Toyota Production System (TPS) [18] or "Lean Production System" named by J.P. Womack et al. [19] of MIT with the soft technology of quality control (SQC and TQM) [20-21].

The author believes that SQC greatly contributed to QCD study activities on manufacturing lines by technical staff in solving engineering problems when Toyota and other automotive manufacturers of Japan made efforts in catching up to the top level manufacturers of the world. As is seen in companies awarded with the Deming Prize, the above-stated fact is universally seen, based on a number of achievements made with the application of SQC [22].

In recent years, however, consumers' sense of value for vehicles has grown more diversified, and it has become indispensable for automotive manufacturers to create high quality, clean and reliable products that appeal to consumer sensitivity. To do this, the engineering capability for designing products that can realize a high degree of reliability must be improved by simultaneously enhancing the level of collaboration activities with parts manufacturers.

Similarly, it is a basic requirement for creating high quality, inexpensive vehicles by applying still more advanced production engineering and production management systems. Through this connection, the most important challenge manufacturers face in the future is how to build high-quality vehicles on assembly lines, with an enhanced rate of operation, by fully utilizing high-precision and high-efficiency production facilities.

As thus far stated, the scope of activity for quality control has expanded from the manufacturing line on the lower stream to development and design on the upper stream. Accordingly, the concept of quality has inevitably changed from product quality- to business process quality orientation. To overcome this challenge, manufacturers must evade conventional quality management approaches that largely depend on individual experience and intuition of key staff, i.e. engineers and management for enhancing technological capability. Instead, the author strongly believes in the necessity of establishing new quality management principles and core technology that comprehensively and effectively utilize the experience and know-how scattered across different levels, divisions and persons of the company.

On such a background, Toyota in recent years has adopted a new quality control principle, Science SQC so called Amasaka SQC Method, which promotes the intellectual productivity of engineering staff and managers to achieve remarkable results from the aspect of a quality strategy [12-13][23]. Reflecting the results, the company has been further promoting TQM-S (TQM by utilizing Science SQC) by positioning it as an important element of next-generation TQM, and the core technique of Toyota and its group's quality strategy, according to a proposal from the author [12][24].

This book provides an overall explanation of the new quality management methodology, Science SQC, and reports how quality strategies based on Science SQC were implemented at Toyota Motor Corporation. This book mainly consists of quotations from the author's theses announced and publicized in Japan and overseas.

2, Science SQC, New Quality Control Principle, includes Chapter 1, SQC Renaissance as an application of the SQC Promotion Cycle; Chapter 2, Scientific SQC Approach, as an application of the new SQC Technical Methods; Chapter 3, "Science SQC" proposal, as an effective implementation of Management SQC; Chapter 4, "Science SQC" Implementation, as stratified SQC training to enhance engineers' problem solving capability; Chapter 5, "Science SQC", A New Quality Control Principle," as a proposal for next-generation TQM, "TQM-S", which played a key role in Toyota's quality strategy.

3 and 4 outline Toyota's quality strategies that utilized "Science SQC". 3, entitled "Human Resource Development and Practical Outcomes of Science SQC in Toyota" includes Chapter 6, SQC Promotion Cycle Activities, for engineers' human resources development and practical results; Chapter 7 outlines user-friendly SQC software, TPOS, that acted as a driving force in the SQC Renaissance; Chapter 8, Management SQC, lead by general managers and managers; and Chapter 9, TSIS-QR System, a sub-core element of SQC integration network.

4 entitled Technological Quality Strategy in Toyota introduces specific studies. 4-1 highlights intellectual product development through the introduction of studies in product plan design, development design, reliability design, and CAE design from Chapters 10 to 13. 4-2 highlights intellectual production technology through introduction of studies in production engineering, process design, production preparation, production process, and process control from Chapters 14 to 18. Finally, 5 provides overall conclusion of the topics covered in this book.

References

[1]For example, Walton. Mary,(1988), The Deming Management Method, *Dodd, Mead & Company, Inc., New York.*
[2]For example, Walter. A. Shewhart,(1986), Statistical Method from the Viewpoint of Quality Control, *Edited and with a New Foreword by W. Edwards Deming, Dover Publications, Inc., New York.*
[3]For example, J. M. Juran,(1989), Juan on Leadership for Quality - An Executive Handbook, *The Free Press, A Division of Macmillan, Inc,.*
[4]K. Amasaka (1999), Special Lecture :The TQM Responsibilities for Industrial

Management in Japan -The Research of Actual TQM Activities for Business Management-, (in Japanese) *Journal of the Japanese Society for Production Management, The10th Annual Technical Conference,* pp.48-54.

[5]For example, T. Goto, (1999), Forgotten Origin of Management -Management Quality Taught by GHQ, (in Japanese) *CSS Management Lecture, Productivity Pub,.*

[6]K. Amasaka et al., (2000), Recommendation of New SQC Training that Assists Businessmen in Job Execution (Part 2) -New Development of QC Education in the Outside Educational Institution- (in Japanese) *Journal of the Japanese for Quality Control, The 30th Technical Conference,* pp.33-36.

[7]Examples: (1) Nikkei Sangyo Shimbun, Strict Evaluation of TQM-Development of Easy-use Techniques,(June 15,1999).(2) Nikkei Sangyo Shimbun, Improving Problem-Finding Capabilities-Reconstruction of a New System to Replace QC,(September 23,2000).(3) Nikkei Sangyo Shimbun, Management Viewpoints-The Collapsing Quality Myth of Japan- (September 24,2000),(4) Nikkei Sangyo Shimbun, Increase in Vehicle Recall up to 40%,(July 6,2000).

[8] K. Amasaka, (1999), TQM as a foundation of Japan (Part 2), TQM Methodology from Management Point of View-Recommendation of New SQC Application- (in Japanese) *Union of Japanese Scientist and Engineers, The 68th Quality Control Symposium,*pp.39-58.

[9] K. Amasaka, (1998), Application of Classification and Related Methods to the SQC Renaissance in Toyota Motor, *Data Science, Classification and Related Methods,* pp. 684-695, *Springer.*

[10] K. Amasaka,(1999), A study on "Science SQC" by Utilizing "Management SQC" -A Demonstrative Study on a New SQC Concept and Procedure in the Manufacturing Industry-, *International Journal of Production Economics,* Vol.60-61, pp.591-598.

[11] K. Amasaka, and S. Osaki,(1999), The Promotion of New Statistical Quality Control Internal Education in Toyota Motor -A Proposal of "Science SQC" for Improving the Principle of TQM-", *The European Journal of Engineering Education,*Vol.24, No.3, pp.259-276 .

[12] K. Amasaka, (2003), Proposal and Implementation of the "Science SQC" Quality Control Principle, *International Journal of Mathematical and Computer Modelling. Modelling,* Vol. 38, No. 11-13, pp. 1125-1136.

{13} Edited by M.Kamio and K.Amasaka,(1992) Cases of SQC Application to Develop Proprietary Techniques (in Japanese), *compiled by Nagoya QC Research*

Group, Japanese Standard Association.

[14] K. Amasaka, (1993),SQC Development and Effects at Toyota, (in Japanese) *Journal of the Japanese Society for Quality Control, Quality,* Vol.23,No.4, pp.47-58.

[15] Edited by K. Amasaka, (2000), Science SQC - Revolution of Business Process Quality (in Japanese), *compiled by Nagoya QST Research Group, Japanese Standard Association.*

[16] K. Amasaka, (2000), Partnering chains as the Platform for Quality Management in Toyota, *Proceedings of the World Conference on Production and Operations Management, Sevilla, Spain,* pp.1-13.

[17] K. Amasaka,(2003), Chapter 6: Quality Management and Supplier of Automobile Industry-Practical Joint Activities between Automobile Manufacturer and Supplier-, (in Japanese), *Changes in Manufacturers' Ordering Systems for Parts and Related Measures - Conditions for Supplier Survival in the Automotive Industry, Serial No. 74, General Research Organization on Medium to Small Enterprises,* pp.113-161.

[18] Examples:(1) T. Oono, (1978), Toyota Production System-Toward Off-Scale Management-, (in Japanese) Diamond-sha. (2)Edited by Toyota Motor Industry, A Glossary of Special Terms Used at Production Job-sites in Toyota (1987) and The Toyota Production System (1995). (in Japanese)

[19] J. P. Womack et al., (1991), The Machine that Change the World, *MIT Press, Harper Prennial, New York.*

[20] K. Amasaka and K. Yamada, (1991), Re-evaluation of the Present QC Concept and Methodology in Auto-Industry, (in Japanese) *Union of Japanese Scientist and Engineers, Total Quality Control,* Vol.42, No.4, pp.13-22.

[21] K. Amasaka and H. Azuma, (1991), The Practice of SQC Training at Toyota - Improving Human Resource Development and Practical Outcomes, (In Japanese) *Quality, Journal of the Japanese Society for Quality Control,* Vol. 21, No. 1, pp. 18-25.

[22] Examples: (1) K. Amasaka and S. Ootani, (1972), Development of Carrier Assembly Inspection Variables for Differential Noise Management, (in Japanese) *Union of Japanese Scientist and Engineers, The 2nd Sensory Inspection Symposium,* pp.5-12. (2) H. Shimizu and K. Amasaka, (1975), Steering Quality Assurance for low speed driving, (in Japanese) *Union of Japanese Scientist and Engineers, Total Quality Control,* Vol.26, No.11, pp.42-46. (3) K. Amasaka, (1983), Mechanization of Work by Knack and Feeling-Straightening of the Rear Axle Shaft-, (in Japanese) Journal of the Japanese for Quality Control, The 26th

Technical Conference,pp.5-10. (4) K. Amasaka and K. Sugawara, (1988), QCD Study Activities by Line Engineering Staff -Improvement of the Corrosion Resistance of Coating for Shock Absorber Parts (in Japanese) Union of Japanese Scientist and Engineers, Total Quality Control, Vol.39, No.11, pp.337-344. (5) K. Amasaka et al., (1990), Corrosion Resistance of Coating for Parts on Chassis of Vehicles-Collaborative QCD Activities, (in Japanese) Coatings Technology, Vol.25, No.6, pp. 230-240.

[23] Examples: (1) Nikkei Mechanical, Special Program-SQC Renaissance in Toyota-Applying Statistical Methods to Routine Engineering Reviews (in Japanese), No.422, pp.24-35 (February 21,1994). (2) Nikkei Mechanical, Close-up, Toyota Group Cooperation under SQC-Achieving Compatibility between Brake Effectiveness and No Brake Noise, (in Japanese) No.552, pp.54-59 (September 8,1998). (3) Nikkei Business, Special Program - Revival of Japanese Quality, Toyota System for World Standard Again-Spreading the Application to Design and New Product Development, (in Japanese) No.931, pp.29-31(March 9,1998).

[24] K. Amasaka, (2000), "TQM-S", A New Principle for TQM Activities: A New Demonstrative Study on "Science SQC", *Proceedings of the International Conference on Production Research, Bangkok, Thailand,* pp. 1-6.

2. Science SQC, New Quality Control Principle

Chapter 1: SQC Renaissance

SQC Promotion Cycle Activities:

SQC Promotion Activities under the Banner of "SQC Renaissance" in Toyota Motor Corporation

To capture the true nature of making products, and in the belief that the best personnel developing is practical research to raise the technological level, we have been engaged in SQC promotion activities under the banner "SQC Renaissance". The aim of SQC promoted by Toyota is to take up the challenge of solving vital technological assignments, and to conduct superior QCDS research by employing SQC in a scientific, recursive manner. To this end, we must build Toyota's technical methods for conducting scientific SQC as a key technology in all stages of the management process, from product planning and development through to manufacturing and sales. Especially, multivariate analysis resolves complex entanglements of cause and effect relationships for both quantitative and qualitative data. A part of application examples are reported below, focusing on the cluster analysis as a representative example.

Keywords: Toyota's SQC Renaissance, SQC Promotion Cycle, Scientific SQC, New SQC Methodology, goodness of invention and patents, preventing vehicle's rusting.

1. Toyota's SQC Renaissance for Making the Most of Staff

These days, a review of the changes in the environment surrounding the manufacturing industry indicated an ever greater necessity for corporate efforts to amplify and capitalize upon the technical prowess of young engineering staff who bear the main burden of these times. To capture the true nature of products, and in the belief that the best personnel developing is practical research to raise the technological level, we have gained a new awareness of Statistical Quality Control (SQC) as a behavioral science. In recent years, the entire company has been engaged in SQC promotion activities as shown in Fig. 1, under the banner "SQC Renaissance" (Amasaka, 1993).

Specifically, the objectives are that all members including the engineering staff and the management should seek to reach excellent solutions for technical problems and that they should realize practical achievement by both improving the proprietary technologies and management technologies through SQC practices. Another aspect of the objective is to develop the SQC promotion cycle activities in which the SQC practices result in practical and full development of SQC education that facilitates effective development of human resources, which, in turn, will be reflected in the performance of operations (Amasaka and Azuma, 1991).

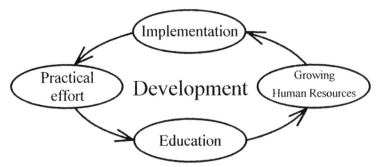

Fig. 1. Schematic Drawing of Company-wide SQC Promotion Cycle Activities

If we are to make products that can satisfy our customers, we must work under the most appropriate conditions for raising work quality and minimizing problems. If staff keep a careful watch over their work and apply SQC properly, SQC can assist them in remedying work processes and effectively raise the quality of work (Amasaka and Yamada, 1991).

2. SQC for improving Technology

The aim of SQC promoted by Toyota is to take up the challenge of solving vital technological assignments, and to conduct superior Quality, Cost and Delivery (QCD) research by employing SQC in a scientific with the exhibition of insight,

- In order to deliver attractive product to customer -

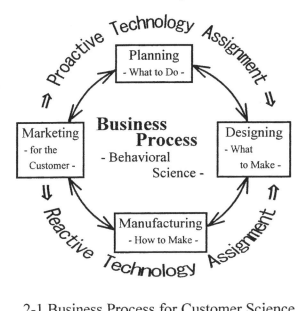

2-1 Business Process for Customer Science

2-2 Scientific SQC for Improving Technology

Fig. 2. New Schematic Drawing of Scientific SQC to Conduct of Superior QCDS Research

inductive problem-solving methodology in addition to engineer's deductive work methods.

This SQC goes beyond reactive technological assignments, to solve proactive technological ones that must be anticipated. This is not a matter of merely performing analytically-oriented SQC in the form of statistical analysis, but is a scientific application of SQC at all stages from problem construction, assignment setting, through to the achievement of objectives, and entails the planning of surveys and experiments to ascertain the desirable scenarios, and devises approaches for tackling problems.

When engineers and managers place value on logical thinking, they can resolve the cause and effect relationships of the gap between theory and practice, and can obtain new facts and knowledge for improving proprietary technology and managerial techniques. Rather than ending up with one-off solutions and partial solutions, they can create technology for general solutions, which in turn leads to improve product quality (Amasaka, 1995).

By operating this way, a wide variety of SQC practical reports should be utilized as guidelines that can contribute to building up of wealth of engineering technologies as well as support for handing down and developing engineering technologies (Kamio and Amasaka, 1992). Based on this view point, we propose new schematic drawing of Scientific SQC that all departments make a superior QCDS Research at each step of business process, as shown in Fig. 2.

3. SQC Established as a Toyota Technical Methodology

In order to be able to provide customer-oriented, attractive products, it is important that we implement customer science that deftly reflects the feelings and voices of customers in the products we make. To this end, we must build Toyota's technical methods for conducting Scientific SQC as a key technology in all stages of the management process, from planning and design through to manufacturing and marketing, as shown in Fig. 2.

3. 1. SQC as a core of the technical methods

Using proprietary technologies and acquired knowledge, SQC resolves complex entanglements of cause and effect relationships for both quantitative and qualitative data. Hence, it is a highly convenient method for technological analyses for improving proprietary technology. New seven tools for Total Quality Control (N7) and basic SQC methods enable full support for designing the experimental process and the analytical process, thus making it possible to analyze technology rapidly and with error-free thinking.

In addition by capitalizing upon proprietary technology, the use of the multivariate analysis method enables 70% to 80% of the mileage required to go before finding solution for a problem to be covered. Combined with SQC methods such as design of experiment, the remaining distance can be covered effectively.

The detailed practical reports shown in the references (Amasaka et al. 1992) (Amasaka et al. 1993) (Amasaka et al. 1994) (Amasaka et al. 1996a) (Amasaka and Sakai, 1996) (Kusune and Amasaka, 1992) (Takaoka and Amasaka, 1991), unravel complex entanglements of cause and effect relationships, by capitalizing on SQC methods such as N7, multivariate analysis and design of experiment in effective combination with the physical and scientific methodology.

The use of SQC methods brings about the outstanding achievement on the jobs, using analysis of sources of variation, modeling for prediction and control (Amasaka et al. 1993) (Amasaka et al., 1994) (Kusune and Amasaka, 1992) (Takaoka and Amasaka, 1991), and concurrent application with neural network as the new technical method (Amasaka et al.,1996a).

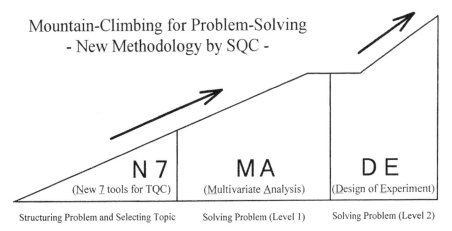

Fig. 3.   Schematic Drawing of Toyota's Technical Methods for Conducting Scientific SQC in Toyota

Moreover, IT (Information Technology) needed for production control (Amasaka and Sakai, 1996), the equipment diagnostic technology (Amasaka et al., 1992) are capitalized on as scientific support. These methodologies for technological problem-solving have been established as new SQC methodology and Toyota's technical method for improving quality works done by engineering staff and managers. Fig. 3 shows a conceptual diagram of new SQC methodology.

3. 2. Using multivariate analysis as the core of Scientific SQC practices

Upstream product technology departments are required to quickly develop advanced design technology. Large amounts of data cannot be obtained from individual R & D projects, but lots of the collected trial and experimental data have excellent features in terms of status and condition control. For this reason, even time-series data collected from similar dead and buried in the R&D projects can be used for gaining scientific insights into the cause and effect relationships through multivariate analysis that makes use of proprietary technology. Thus we are able to build logical quantitative models for analysis of source of variation, prediction and control (Amasaka et al., 1996a) (Takaoka and Amasaka, 1991).

Midstream manufacturing preparation departments are also required to rapidly develop new production technology and control systems. Comparatively large amounts of data can be obtained, although they stem from R & D, trials, and attempts using mass-production facilities, and rather than being planned, they tend to be of the trial and error variety, changing over time series.

The soundness of these data is not as clear as those of data from upstream departments, but they are data gathered on the spot by engineers, and are reliable in terms of the status and conditions pertaining at the time of collection. Hence, they lend themselves to the application of multivariate analysis using proprietary technology, so that technological knowledge and new facts can be probed for use in raising the technological level (Amasaka and Sakai, 1996) (Kusune and Amasaka, 1992).

In the downstream manufacturing departments, the important task is to improve manufacturing technology for making stable, high-quality products. Although the reliability of the collected data is not as clear as for the upstream and midstream departments, a great deal of raw data can be obtained if they are collected in a purposeful and planned fashion.

Then, by taking advantage of the manufacturing engineers' unique insights into the actual status of the site and the product, multivariate analysis can be performed to analysis of sources of variation, to extract factors that may aid in quality improvement, and to control processes at the optimum level, thus contributing to improvement of process capability (Amasaka et al. 1992) (Amasaka et al. 1993) (Amasaka et al., 1994).

In this way, multivariate analysis can be used for flexible analysis of technology even with varying quantities of diverse data collected in the past. It has taken root not only as an analytic method of SQC specialists, but also as Toyota's technical methods for skillfully raising the work quality of engineers (Amasaka et al. 1996b) (Amasaka et al., 1996c) (Amasaka and Kosugi, 1991) (Amasaka and Maki, 1991).

Recently, we have developed and provided our staff with friendly SQC analysis software to be used from personal computers (Amasaka et al., 1995a) (Amasaka and Maki, 1992). In all types of technological fields, multivariate analysis has become the core of the SQC practices in applying scientific use of the technical field, and practical research is progressing. A part of application examples are reported below, focusing on the cluster analysis as a representative example.

4. Examples of Applying Multivariate Analysis Method Focusing on Cluster Analysis

The research examples outlined are "Analysis of sources of variation in vehicle rusting", both belonging to the product and production technology arenas, and "Latent structure of engineers' attitudes to good inventions and patents", a theme impinging on the technological development and control areas. All these research examples started out with cluster analysis to unravel the complex technical assignments, and show that the application of a combination of different multivariate analysis methods has brought about the expected results.

4. 1.　"Latent structure of engineers' attitudes to goodness of inventions and patents"

Acquiring "good patents" that enable intelligent properties in possession to improve the corporate quality, has become more important measures for allowing permanent business activities.

Hence, good patents recognized by managers and technical staff (here-in-after referred to as engineers), containing the contents of both invention and right, are clarified in terms of quality and required to conduct researches aimed of encouraging the acquisition for strong and wide patents. Therefore, this paper selects the representative examples that grasp the conceptual structure of the "good patents" admitted by the engineers (Amasaka et al., 1996d) (Amasaka and Ihara, 1996) (Ihara and Amasaka, 1996).

4. 1. 1. Characteristic Classification of Common Images of "Good Patents" and Grasping their Structural Concept

A good patent is said that the invention and right leads to the benefit of own company while allowing the company to maintain its main business and affecting others . The patent information available currently gives only simple statistical values, not studied about qualitative analysis by probing into the conscious region of engineers as inventors. In this connection, the survey aimed of objectifying the subjective structural concept of "good patent" is employed by grasping the engineers' latent status-quo recognition interpreted as the language information.

Regarding to the survey's process (Amasaka et al., 1995b), free opinions are collected from engineers in advance about "what a good patent is, while considering the recent environmental situations." Those collected opinions are grouped and arranged into the key words by employing an affinity chart method and a relation chart method in the cooperation with the special department for the

Table 1. Example: "What is a good patent?"

● Please answer the questionnaire for constructing patent strategy. These days, patent disputes with other companies are increasing. Under the circumstance, how do you think about " What is a good patent ? " ※ Encircle the number.	《Assessment》 1: very important 2: rather important 3: important 4: either 5: bit unnecessary 6: unnecessary

[1]   Question about "inventive technic."   [2]   Question about "patent right."

(1) practicality   `1 2 3 4 5 6`   (13) large system   `1 2 3 4 5 6`

(2) cost   `1 2 3 4 5 6`   (14) many examples   `1 2 3 4 5 6`

(3) commercialization   `1 2 3 4 5 6`

patent application.

Then the key words are largely classified into "content of inventive technique" and "content of patent right." Respective contents are summarized into a format of 11 questions as outlined in Table 1 sheet. The survey by the questionnaire in marking the Table 1 sheet, one of given answers for selection, is conducted to a total of 97 persons selected from among those with experience of patent acquisition in seven departments from Product Technology: Research "b", Design "c", Development "d", Technical Administration "a", Production Engineering "e", "f", and "g", as shown in Fig. 4.

Fig. 4. 1 shows the grouping by shared recognition of the "good patent" through the cluster analysis. The results are classified mainly into three major clusters ((1), (2), and (3)). The figure shows the percentage of the total number of persons who answer the questionnaire and those of each cluster of the seven departments. The figure indicates that the percentage is dispersed by the department. The questionnaire results are now analyzed with factor analysis by probing into the commonly shared structural concept of the "good patent." Fig. 4. 2 shows the result of the analysis obtained from the scatter diagram of factor loading. Varimax method allows the reading of the strength of the right and the engineering development capability standing opposite to each other on axis 1 and the commercialization and profit-minded on axis 2.

The result of this analysis confirms that there exists in the seven departments four types of engineers who attach importance to commercialization, profit-minded, the strength of the right, or the engineering development capability. Analysis of the similar scatter diagram for the factor scores reveals an interesting new knowledge that each engineer has his own structural concept of the "good patent."

And the summaries of these two analytical results allow us to interpret, as shown in the lower section of Fig. 4. 1, that group (1) is A: the realist group (that places importance on practical invention based on Toyota's engineering capability and on the right of practical effect); group (2) is B: the advance group (that places importance on advanced invention superior to competitors or specific right competitors are eager to get); and group (3) is C: the futuristic group (that places importance on the prospective invention on the construction stage and the right to lead competitors with internationally).

For example, research section "b" mostly consists of the futuristic group while Technical Administration "a" is largely composed of the realist group and so on for the rest of sections that reflect logical result of analysis. Reganding to this

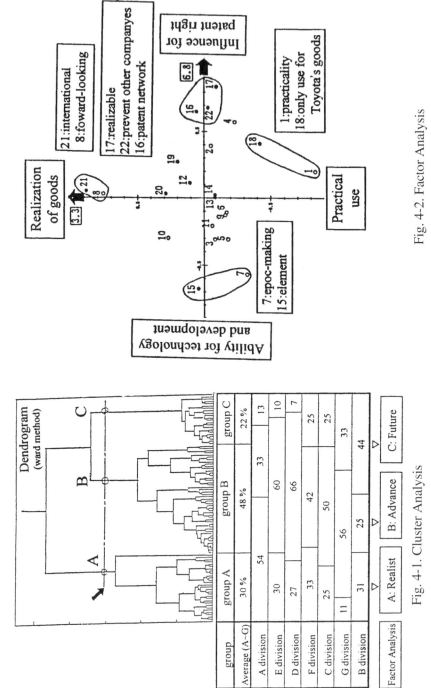

Fig. 4-2. Factor Analysis

Fig. 4-1. Cluster Analysis

Fig. 4. Patterning and common understanding of good patents

respective departments are conscious and in need of well-balanced activities for "good patent" as they understand what to expect from and what role to play in executing their job. We have thus obtained valuable results (of the analysis) which constitute the preparation for our future patent strategy.

4. 1. 2. Grasping Linkage of Recognition between "Good Invention" and "Good Right"

As mentioned in the proceeding section, a "good patent" has two sides of invented technique and subsequent right. To determine characteristic key words for respective sides as recognized by the realist, advance, and futuristic groups classified as the results of cluster and factor analyses, and obtain the degrees of mutual linkage among them, canonical correlation analysis is conducted.

Fig. 5 shows the result of the analysis. It is found from the main component relation scatter diagram in the figure that most of the engineers belonging to A: the realist group can be related to the profit-minded in terms of the content of invented art and to the effective-minded in terms of the content of the right by the key words that each group is required to have.

Likewise, the figure reads that the B: the advance group and C: the futuristic group are configured with the pioneer-minded and the innovative-minded, and the competitive-minded and the monopoly-minded respectively. This result of analysis can be judged logicalty from the viewpoint of empirical technique.

We are able to obtain a verification to the effect that such cause and effect relationships can be a qualitative general evaluation index for a patent application by weighting respective key words as the multiple regression equation. At present, this process is in the middle of transfer to the implementation stage as a regular task. This case (1) shows a scientific application of SQC exactly as seen in Fig. 2 by upgrading the quality of job on the business process stage in a proactive engineering area. It is thus judged that the application effect of multivariate analysis as the core of Toyota Technical Method has been verified.

4. 2. "Analysis of sources of variation for preventing vehicles' rusting"

One of the subjects for technological development of higher quality, longer life vehicles is the quality assurance of anti-rusting of the body. From the engineering

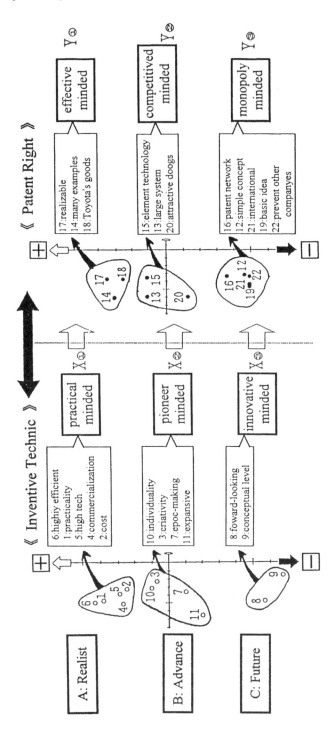

Fig. 5. The Relate of Inventive Technic and Patent Right (canonical correlation analysis)

viewpoint of anti-rusting of vehicle body, it involves consideration for structural design, adoption of anti-rusting steel, adoption of local anti-rusting processing (such as the application of wax, sealer, etc.), and paint design (conversion treatment and improvement of electrodeposition coating, etc.).

On the implementation stage of the research, these anti-rusting measures are adopted singularly or in combination of multiple measures as may be required by the construction of subject section or corrosion factors. In order for us to proceed with advanced and timely QCDS study activities, it is necessary to conduct variable factor analysis of vehicle's anti-rusting and the optimization. In this sense, SQC centering around the multivariate analysis has much to contribute as a scientific approach. In this connection, this paper describes a characteristic case for study (Amasaka et al. 1995c).

4. 2. 1. Anti-rusting Methods and How to Outline their Characteristics

To establish a superior quality assurance system, it is important to set up the network of quality assurance on the job processing stage so as to raise reliability technology of whole the sectors including product planning, design, review, production engineering, process design, administration, inspection and so on. Such a quality assurance activity under the cooperation of all the sectors has been established as Toyota's QA network, where SQC plays an important role as the behavioral science for enhancing quality performance.

Toyota has been incorporating various anti-rusting methods to various sections of the vehicle. To outline the deployment of quality performance, the application of matrix diagram method is effective. For example, Table 2 outlines complex correlationship, between corrosive environmental factors of 72 sections of a vehicle divided for the ease of arrangement of anti-rusting measures and the manufacturing processes factors, and the respective anti-rusting methods adopted in an anti-rusting QA network table. This table is a summary of objective facts inductively arrested by engineers from multiple engineering sectors, which are then arranged deductively in the subjective point of view.

To make this table more effective, it is necessary to provide it with the ease of visual recognition so that it shows at a glance where Toyota stands in its present activities for rust prevention quality assurance. Table 2 enables staff and engineers to make the engineering judgment more accurately, subsequently contributing much to the strategic decision making on the part of the

Table 2. Anti-corrosion Q. A. Matrix

| Part | | | Category Rest Prevent | | | | |
Large Classificati	Small Classification	No	Electoro-deposited Coating	Corrosion Prevented Steel Sheet	Adhesive	Wc S
Heming of Shell Parts	Lower part of door	1		1	1	
	Lower part of laggage	2		1	1	
	Lower part of fuel filler lid	3		1		
	Door, The others	4		1	1	
Shell Parts * R/F	Hood * Lock R/F	5	1	1		
	Door * Lock R/F	6		1		
	Door * Side Protection Bar	7	1	1		
	Back Door * Lock R/F	8		1		
Front Floor	General Surface	9		1		
	Under R/F	10		1		
	Exhaust Pipe R/F					

↓ 72 parts   28 categories
   Every part of   Anti corrosion factor
   body shell   Process influence factor } Yes: 1, No: blank
                    Corrosion factor   etc.

management. To proceed with such an aim and make it much easier to understand complex entanglements of cause and effect relationships, they will be summarized visually as shown in Fig. 6 by using quantification method type III. From a scatter diagram of vehicle section (axis I x axis II), it is apparent that Toyota's anti-rusting measures are taken mainly against the steel sheet joint portions and that the local anti-rusting processing and many other methods tend to be adopted as the measures reach the underbody sections. The diagram indicates that the combination of several types of anti-rusting methods is applied to the doors and underbody members positioned in the upperbody sections.

  Adoption of such an analytical method makes it easy to evaluate the vehicles of competition, enabling the reactive bench marking for additional advantages. In addition, insight into proprietary technologies with the knowledge acquired from the result of this analysis enables us to grasp proactively our responses to the future quality assurance including the trends of anti-rusting measures and techniques of competition that reflect thethe target quality and the counter-marketability of anti-rusting materials.

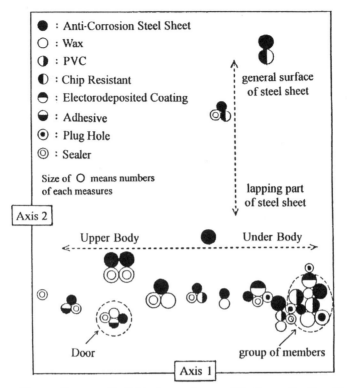

● : Anti-Corrosion Steel Sheet
○ : Wax
◑ : PVC
◐ : Chip Resistant
◒ : Electorodeposited Coating
◓ : Adhesive
◉ : Plug Hole
◎ : Sealer

Size of ○ means numbers
of each measures

Axis 2

general surface
of steel sheet

lapping part
of steel sheet

Upper Body   Under Body

Door   group of members

Axis 1

(Scatter diagram of individual-score by quantification method type3)

Fig. 6. Characteristic of Toyota's Anti-Corrosion Measures

4. 2. 2. Factorial Analysis Method for an Optimal Anti-Rusting of Vehicle Bodies

This subsection takes up the door hemming sections to which application of anti-rusting technique meets particular difficulties. Spraying of snow melting salts during winter to prevent the roads from freezing allows the filtration of salt water, a corrosive factor, to the hemmed joint portions of door outer-panels and inner-panels of rust prevention steel sheet. To prevent this, wax and/or sealer are adopted or the joint portions of the door outer-panels and inner-panels are sealed with adhesive agent for the dual purpose of adhesion and anti-rusting.

To evaluate the performance of various anti-rusting measures, we conduct a market monitor test using actual vehicles, in-house accelerated corrosion test, and the accelerated corrosion test on the bench using testpieces and/or parts. For any of these tests, it is imperative to conduct analysis of sources of variation for the optimization of anti-rusting of the body.

Table 3. Experiment Data of Test Piece

Anti-corrosion specification (combination)* and Result

Factor and Level					Corrosion depth	
X1	X2	X3	X4	X5	outer	inner
1	1	2	2	2	0.15	0
1	1	1	2	2	0.10	0
2	1	2	1			

X1: type of steel (outer)   1: steel sheet

                                2: anti corrosion steel sheet level 1

X2: type of steel (inner)   1: anti corrosion steel sheet level 1

                                2: anti corrosion steel sheet level 2

X3: spot anti corrosion steel process A (1:no 2:yes)

X4: spot anti corrosion steel process B (1:no 2:yes)

X5: spot anti corrosion steel process C (1:no 2:yes)

*: 21 combinations

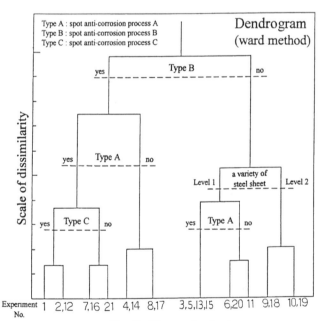

Fig. 7. Cluster Analysis

Example indicated in Table 3 outlines the test data showing the result of analysis on the relationship between the (combined) rust prevention specifications of a testpiece and the depth of corrosion. A dendrogram can be generated as shown in Fig. 7 by subjecting the above data to a cluster analysis. This dendrogram is used to hierarchically outline the degree of effect from the factors of anti-rusting measures by grouping the experiment numbers of a corrosion test.

Moreover, analysis using quantification method type I enables us to verify quantitative degree of effect from the factors of anti-rusting measures by category-scores and the size of partial correlation coefficient, as shown in Fig.8. From this diagram, it is possible to grasp the effect of anti-rusting steel sheets used to the door hemming sections and the validity of quantitative effect of local anti-rusting processing A, B, and C from the engineering point of view, which in concurrent application with the analytical results of other testing methods using actual vehicles enables the optimization of the vehicle body rust prevention.

We have applied comprehensive and technical insight into these results of analyses. And by adopting controllable factors effective for anti-rusting measures in addition to the environmental factors in consideration of market environments, we have been able to verify deftly the factorial effects with the application of the design of experiment. We have thus succeeded in the embodiment of production with good QCDS performance through the optimization of the vehicle body rust prevention.

It is judged that this case (2) too constitutes a new methodology of mountain-climbing for Problem-Solving effectively using SQC in concurrent application of multivariate analysis with N7 and design of experiment. We think that here too it has been verified that the multivariate analytical method forms the core of the Toyota Technical Method for scientific implementation of SQC as observed in Fig. 2 and 3.

5. Conclusion

We have been able to verify the following from the two demonstrative cases for studies as above mentioned:In combination with N7 and design of experiment, various multivariate analysis starting from the cluster analysis have been established as the core of technology-advancing SQC from a mere statistical analysis that tends to place unbalanced importance on analysis.

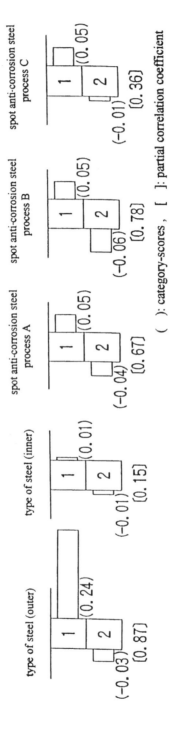

Fig. 8. Influence of Anti-corrosion Measures by Quantification Method (type 1)

We think that we have verified that the multivariate analysis offers a great application effect as the core of the SQC methods in connection with the construction of the presently advocated Toyota Technical Method for the scientific implementation of SQC and the enhancement of the effectiveness of called SQC.

In unraveling confound situations and complex entanglements of cause and effect relationships, one of the multivariate analysis' methods cluster analysis enables them to clarify and organize visually and logically. On this view point, engineers' abilities with latent pursuit-minded and new ideas-minded are enhanced and improved. It is considered that this new technological approach has great insight to unravel and pursue complex problem-assignments appropriately. The author appreciates valuable teaching and comments from those people concerned.

References

Amasaka, K. (1993), "SQC Development and Effects at TOYOTA," (in Japanese) *QUALITY, Journal of the Japanese Society for Quality Control*, 23(4), 47-58.

Amasaka, K. (1995), "A Construction of SQC Intelligence System for Quick Registration and Retrieval Library, - A Visualized SQC Report for Technical Wealth -," *Lecture Notes in Economics and Mathematical Systems*, 445, 318-336, *Springer*.

Amasaka, K. and Azuma, H. (1991), "The Practice SQC Education at TOYOTA, - For Growing Human Resource and Practical Effort -," (in Japanese), *QUALITY, Journal of the Japanese Society for Quality Control*, 21(1), 18-25.

Amasaka, K. and Ihara, M. (1996), "Latent Structure of Goodness-of-invention," *Lecture Notes in Economics and Mathematical Systems*, 445, 348-353, *Springer*.

Amasaka, K. and Kosugi, T. (1991), "Application and Effects of Multivariate Analysis in TOYOTA," (in Japanese), *The Behavior Metric Society of Japan, The 19th Annual Conference*, 178-183.

Amasaka, K. and Maki, K. (1991), "Application of Multivariate Analysis for the Attraction of Manufacturing Vehicles," (in Japanese), *The Behavior Metric*

Society of Japan, The 19th Annual Conference, 190-195.

Amasaka, K. and Maki, K. (1992), "Application of SQC Analysis Soft in Toyota," (in Japanese), *QUALITY, Journal of the Japanese Society for Quality Control*, 22(2), 79-85.

Amasaka, K. and Sakai, H. (1996), "Improving the Reliability of Body Assembly Line Equipment," *The International Journal of Reliability and Safety Engineering*, 3(1), 11-24.

Amasaka, K. and Yamada, K. (1991), "Re-evaluation of Present QC concept and Methodology in Autoindustry, - Deployment of SQC Renaissance in Toyota -," (in Japanese), *Total Quality Control*, 42(4), 13-22.

Amasaka, K. et al.(1992), "A Method on Equipment Diagnosis of Grinder," (in Japanese) *QUALITY, Journal of the Japanese Society for Quality Control, The 42th Technical Conference*, 37-40.

Amasaka, K. et al.(1993), "A Study of Quality Assurance to Protect Plating Pants from Corrosion by SQC, - Improvement of Grenading Roughness for Rod Piston by Centerless Grinding -," (in Japanese) *QUALITY, Journal of the Japanese Society for Quality Control*, 23(2), 90-98.

Amasaka, K. et al. (1994), "Consideration of effieientical counter measure method for Foundry, - Adaptability of defects control to Casting Iron Cylinder Block -," (in Japanese), *Journal of the Japanese Society for Quality Control, The 47th Technical Conference*, 60-65.

Amasaka, K. et al. (1995a), " Aiming at Statistical Package using in the Job Process," (in Japanese), *Journal of the Japanese Society for Quality Control, The 25th Annual Technical Conference*, 3-6.

Amasaka, K. et al. (1995b), "A study of Questionnaire Analysis of the Free Opinion, -The Analysis of Information Expressed in Words Using N7 and Multivariate Analysis Together-," (in Japanese), *Journal of the Japanese Society for Quality Control, The 50th Technical Conference*, 43-46.

Amasaka, K. et al. (1995c), "The Q. A. Network Activity for Prevent Rusting of Vehicle by Using SQC," (in Japanese), *Journal of the Japanese Society for Quality Control, The 50th Technical Conference*, 35-38.

Amasaka, K. et al. (1996a), "A Study on Estimating Vehicle Aerodynamics of Lift, - Combining the Usage of Neural Networks and Multivariate Analysis -," (in Japanese), *Journal Institute of Systems Control and Information Engineers*, 9 (5), 229-237.

Amasaka, K. et al. (1996b), "Influence of Multicollinearity and Proposal of New Method of Variable Selection, -A Study of Applied Multiple Regression Analysis for Analysis of Source of Valuation-," (in Japanese), *Japan Industrial Management Association*, 46(6), 573-584.

Amasaka, K. et al.(1996c), "A Study on Validity of the BN method for Variable Selection, - A Study of Applied Multiple Regression Analysis for Analyzing Source of Variation Factors (Part II) -," (in Japanese), *Japan Industrial Management Association*, 47(4), 248-256.

Amasaka, K. et al. (1996d), "An Investigation of engineers' Recognition and Feelings about Good Patens by New SQC Method,"(in Japanese), *Journal of the Japanese Society for Quality Control, The 52th Technical Conference*, 17-24.

Ihara, M. and Amasaka, K. (1996), "Factor Analysis for Selected Observations," *Lecture Notes in Economics and Mathematical Systems*,445, 354-361, Springer.

Kamio, M. and Amasaka, K. (1992), "Collection of Activity Example Using SQC Method to Improve Engineering Technologies," (in Japanese), *Japanese Standards Association, NAGOYA QC Research Group*.

Kusune, K. and Amasaka, K. (1992), "The Statistical Analysis of the Springback for Stamping Parts with longitudinal Curvature," (in Japanese), *QUALITY, Journal of the Japanese Society for Quality Control*, 22(4), 24-30.

Takaoka, T. and Amasaka, K. (1991), "Derivation of Statistical Equation for Fuel Consumption in S. I. Enginenes," (in Japanese), *QUALITY, Journal of the Japanese Society for Quality Control*, 21(1), 64-69.

Chapter 2: Scientific SQC Approach

Utilizing SQC Technical Methods:

A New SQC Concept and Procedure in the
Manufacturing Industry

A review of the changes in the environment surrounding the manufacturing industry indicated an ever greater necessity for corporate efforts to amplify and capitalize upon the technical prowess of engineering staff who bear the main burden of these times. In this study, we propose establishment of a new technical method for conducting scientific SQC as a new concept and procedure that will improve the job quality of engineering staff in each stage of business process so as to contribute to production with excellent QCDS. We also try to evaluate and investigate its usefulness on the basis of an actual study example in Toyota.

Keywords: Manufacturing industry, A New SQC concept and procedure, Scientific SQC, Multivariate analysis.

1. Introduction

Recently a review of the changes in the environment surrounding the manufacturing industry indicated an ever greater necessity for corporate efforts to amplify and capitalize upon the technical prowess of engineering staff who bear the main burden of these times.

In other words, we have to practice customer-oriented behavioral science for producing attractive products. It is necessary for us to contribute to creation of technologies by making scientific use of SQC (Statistical Quality Control) as behavioral science for proactive solution of important technological problems in addition to reactive solution of engineering problems.

In this study, therefore, we propose establishment of a new technical method for conducting scientific SQC as a new concept and procedure that will improve the job quality of engineering staff in each stage of business process so as to contribute to production with excellent QCDS (quality, cost, delivery and amount, safety and satisfaction). We also try evaluation and investigation of its usefulness on the basis of an actual study example in Toyota.

2. A Proposal of a New SQC Concept and Procedure

If we are to make products that can satisfy our customers, we must work under the most appropriate conditions for raising work quality and minimizing problems. If engineering staff keep a careful watch over their work and apply SQC properly, SQC can assist them in remedying work processes and effectively raise the job quality.

One can use the SQC method for the improvement of technological development and to build up the technological assets, SQC methods are utilized not as an analytical tool by SQC specialists , but have been studied systematically and established as the methodology of an empirical science for improving technology.

In this paper, we propose the establishment of Toyota's new technological method as a scientific SQC practice, and we give some consideration to the usefulness of this new SQC concept and procedure in the manufacturing industry. [1]

2.1. Scientific SQC for Improving Technology

In order to be able to provide customer-oriented, attractive products, it is important that we include customers' opinions that deftly reflect the feelings and voices of customers in the products we produce. Consequently, we have to do practical research to raise the technical level of capturing the true nature of products. This has to give a new awareness to the importance of SQC as a behavioral science in all stages of the business process, from planning and design through to manufacturing and marketing.

The aim of SQC promoted by Toyota is to take up the challenge of solving vital technological assignments by all members including the engineering staff and management, and to conduct superior QCDS research with the exhibition of insight by employing SQC in a scientific, inductive problem-solving methodology in addition to engineer's deductive work methods.

This SQC goes beyond reactive technological assignments to solve proactive technological ones that must be anticipated. This is not a matter of merely performing analytically-oriented SQC in the form of statistical analysis, but is a scientific application of SQC at all stages from problem-construction and assignment setting the achievement of objectives. It entails the planning of surveys and experiments to ascertain the desirable scenarios, and devises approaches for tackling problems.

When engineers and managers place values on logical thinking, they can resolve the cause and effect relationships of the gap between theory and practice, and can obtain new facts and knowledge for improving proprietary technology and managerial techniques. Rather than ending up with one-off solutions and partial solutions, they can find general solutions of technology, which in turn lead to improve product quality.

By operating this way, wide varieties of SQC practical reports should be utilized as guidelines that can contribute to building up of wealth of engineering technologies as well as support for handing down and developing engineering technologies [2]. Using the wealth of intellectual and useful knowledge, we can establish the concept of vital technological assignments, and develop the cycle to improve the quality of business process. Based on this view point, we propose a

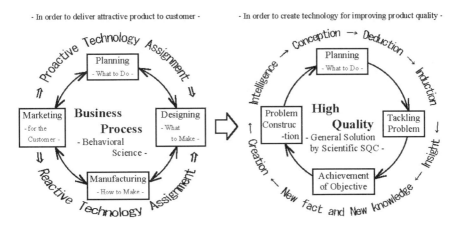

1-1 Business Process for Customer Science   1-2 Scientific SQC for Improving Technology

Fig. 1. New schematic drawing of scientific SQC to conduct superior QCDS research

new schematic drawing of scientific SQC that all departments make a superior QCDS research at each step of business process, as shown in Fig. 1.

2.2. Eatablishment of a New Technical Method for Conducting Scientific SQC

Using proprietary technologies and acquired knowledge, SQC resolves complex entanglements of cause and effect relationships for both quantitative and qualitative data. Hence, it is a highly convenient method for technological analyses for improving proprietary technology. N7 (new seven tools for TQC-total quality control) and basic SQC methods enable full support for designing the experimental process and the analytical process, thus making it possible to analyze technology rapidly and with error-free thinking.

In addition by capitalizing upon proprietary technology, the use of the multivariate analysis method enables 70% to 80% of the mileage required to go before finding solution for a problem to be covered. Combined with SQC methods such as design of experiment, the remaining distance can be covered effectively.

The detailed practical reports shown in the references [3, 4], unravel complex entanglements of cause and effect relationships, by capitalizing on SQC methods such as N7, multivariate analysis and design of experiment in effective combination with the physical and scientific methodology. The use of SQC methods brings about expected technical results, using analysis of sources of variation, modeling for prediction and control, and concurrent application with neural networks as the new technical method. Moreover, IT (Information Technology) needed for production control and the equipment diagnostic technology are capitalized on as scientific support.

Above all, multivariate analysis, which allows flexible technical analysis of various data collected in the past, has become more widely used not only as an analytical method by SQC specialists, but also as (a) SQC technical methods by the engineers in the upstream production engineering department requiring quick development of advanced design technologies, (b) in the production preparation department in the middle reaches requiring efficient development of new production engineering and control technologies, and (c) in the downstream manufacturing department, requiring stabilized high-quality production by improving the production technologies [5, 6].

Recently, we have developed and provided our staff with friendly SQC analysis

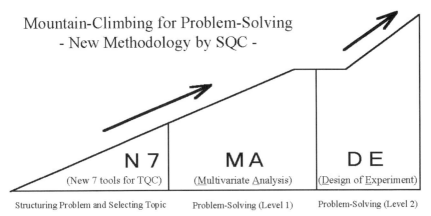

Fig. 2. Schematic drawing of SQC technical methods for conducting
scientific SQC at Toyota

software to be used on personal computers.[7, 8] In all types of technological fields, multivariate analysis has become a core of the SQC practices in applying scientific use of the technical field, so practical research is progressing.[9, 10]

These methodologies for technological problem-solving have been established as a new SQC methodology and Toyota's technical method for improving quality works done by engineering staff and managers. Fig. 2 shows a conceptual diagram of a new SQC methodology.

2.3. Systematic Promotion and Integrated Network System of SQC Activities

In order to be able to proceed in SQC practice systematically, it is necessary for the entire company to have been engaged SQC promotion cycle activities. As detailed in Reference [11], the objectives are that all members including the engineering staff and management, should seek to reach excellent solutions for technical problems, and that they should realize practical achievement by improving both the proprietary technologies and management technologies through SQC practices. Another aspect of the objectives is to develop the SQC promotion cycle activities in which the SQC practices result in practical and full development of SQC education. This facilitates the effective development of human resources, which, in turn, will be reflected in the performance of operations.

SQC application is effective in actual jobs as introduced by various examples in

Section 2. 2. Systematization of SQC application is necessary for contribution to forming technological assets to be inherited and developed as important key technologies for perpetual development of the SQC promotion cycle. The TTIS (Total SQC Technical Intelligence System) shown in Fig. 3 has been constructed as an integrated SQC network system for powerful support of technical analyses by the engineering staff. The main objective of TTIS is support engineers and managers in solving important engineering problems, and to make judgments and decisions explicitly scientific rather than implicitly as in the past. As detailed in references [12-14], the TTIS is an integration of the four main systems shown in the figure so that they will grow together through complementing one another, on assumption that creative production requires a network of high-level engineering force.

As detailed in Fig. 3 [13], the TSIS (Total SQC Intelligence System) can be referred to as a library of SQC application examples that has been constructed as

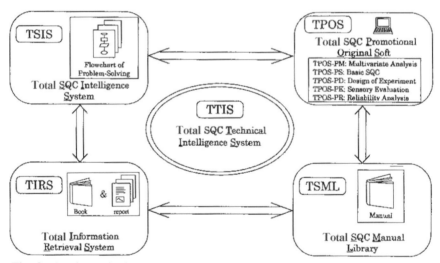

Fig. 3. A schematic drawing of TTIS for SQC information synthesis networks.

an entity consisting of four subsystems for facilitating accumulation of technological assets. For example, the TSIS-QR (-Quick Registration and Retrieval Library) is a visualized professional procedure obtained by pigeonholing the flow of problem solution in steps as a flowchart on one A4 sheet on which new facts and knowledge obtained from amalgamation of technologies and SQC.

It is further networked with the TSIS-RB (-Reference Book), -PM (-Practice Manual) and -ML (-Mapping Library) that describe the recommended newest and best work procedures for classified engineering problems, the summary by engineering field, and the list of SQC applications that contribute to the inheritance and development of technologies, respectively.

The TPOS, as detailed in references [7, 8], is an SQC software package for a personal computer that promotes spiraling the SQC cycle developed within the company upward. The TSML (Total SQC Manual Library) is a library of classified technical methods for scientific application of SQC techniques along the job flow, and the TIRS (Total Information Retrieval System) consists of the technical reports and engineering books enabling the degree of contribution of SQC to be confirmed according to the daily job results.

3. Examples of Applying Technical Methods Focusing on Multivariate Analysis

The research examples outlined are "A problem-solving for the technical bottleneck of engine pinhole" belonging to the reactive technology assignment, and "Study of psychographics for vehicle's profile design and user", a theme impinging on the proactive technology assignment. All these researches examples show that the fundamental and application of technical methods by utilizing SQC to unravel the complex technical assignment has brought about the outstanding achievement on the job.

3.1. Application to Engine Cylinder Block Casting Technologies

Recent engines are becoming complicated in shape for satisfying the exhaust emission control and improving the fuel economy. As a result, molten iron cannot flow smoothly in the sand mold so as to causing small pinholes (cavities) to be generated here and there. Possible causes of this defect have so far been known implicitly, resulting in a bottleneck of casting technologies that hinders permanent correction.

Over 10% of some cylinder block, pinhole defects were found in the later machining process. A casting technology specialist tackled this problem for one year in a trial and error manner, but no appreciable improvement was made in casting technologies. A task team mainly consisting of managers and field engineers well versed in SQC from various sections was organized to tackle this with the problem-solving in two steps[15]. In the first step, a plan was set up to reduce the defect by 1/2 in a short period of half month. The method was for efficient mountain-climbing for problem-solving, without stopping the mass production line, to grasp the causal relationships of quality dispersion as they are without planning any special experiment.

Fig. 4 shows the job flow. First, the pinhole generation mechanism was technically checked orderly by using the chart methods and relation chart method in the N7 technique, and unknown causal relationships were picked up without omission by using the matrix diagram method in the order of processes. Then, the items to be investigated, such as the degree of influence of each pinhole generation cause, were screened by using the cause and system diagram method, and PDPC (process decision program chart) was applied for smooth job execution according to the planned schedule. As a result, necessary corrective actions were sorted out quickly by checking the technical appropriateness of the causal relationships in three days by efficiently collecting the manufacturing condition dispersion data, by stratifying the effects of causes of variation through cluster analysis using SQC software for multivariate analysis as shown in the figure, and by assessing the degree of influence of each cause using principal component analysis and multiple regression analysis.

As shown in Fig. 5 exemplifying the result of analysis source of variation factors, major causes of pinhole defect are casting sand temperature and composition, molten iron temperature, etc. The casting sand temperature is especially highly influential. The mechanism of pinhole defect was able to be estimated by deep analysis using the partial regression plot method, for example, and the result was tested by photographic observation around the defective portion. The processes have been improved accordingly for control to the optimum conditions, and the defect frequency has been reduced to 1/3 as expected.

For further reduction by another half in the second step, the design of experiment was applied to analysis source of variation by setting new control factors and extrapolating the range of factor level. As a result, optimum casting conditions were able to be obtained efficiently through acquisition of new facts and knowledge, for example, by grasping the degrees of influence of factors not regarded as important in the past. By resultant improvements of production technologies, the pinhole defects were reduced to 1/3 as predicted, thereby improving and stabilizing the casting technologies.

In this study example, mountain-climbing for problem-solving was executed in a

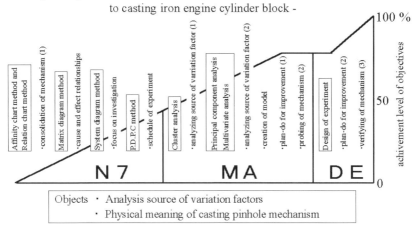

Fig. 4. Representation example by SQC technical methods

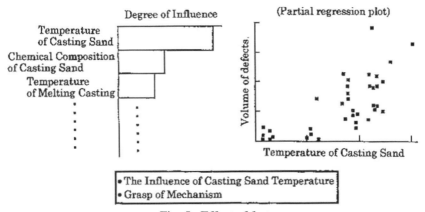

Fig. 5. Effect of factors

short period by the scientific approach as shown in Fig. 1. The study is judged positive as it brought about an excellent engineering result according to the fundamentals of the technical methods shown in Fig. 2. Consolidating the inherent technologies of individual engineers by the team activity as a methodology for solving the technological bottleneck, use of SQC in combination to make implicit knowledge to be explicit, and improving the design of experiment and analyzing to avoid the trial and error approach can be regarded as a positive proof of the new SQC concept and procedures in the manufacturing industry. As detailed in references [16, 17], we have applied this problem-solving approach recently for "improvement of body antirusting" and "improvement of brake squeak" as bottleneck technologies in the engineering and design departments, and have obtained similar results to show its universal applicability.

3.2. A Study of Psychographics for Vehicles' Profile Design and User

Studying the "styles of selling vehicles in the future" is important in design strategy. One characteristic study example is to explore the vehicle value by means of the language of image, with vehicle body style designers as a panel.

Recreational vehicles are selling well recently in Japan. Predicting the styles of selling vehicles in the future is important in the design strategy. We have, therefore, formed a specialist panel consisting of male and female employees, both young and old, in the design department having sufficient knowledge on

Table 1. Images of vehicles

Group	Language of Image	CELSIOR	CROWN	...	CORONA	RAV4	...	BENZ	BMW	...
Status	1. a successful man	34	28		0	0		27	21	
	2. an industrialist	27	27		0	0		29	22	
	3. modest	0	3	...	24	0	...	1	1	...
	4. within one's means	4	4		26	2		2	2	
	⋮	⋮	⋮		⋮	⋮		⋮	⋮	
Life-style	15. traditional	6	29		8	0		28	11	
	16. trendy	6	1		0	14		5	4	
	17. familism	0	3	...	24	0	...	3	1	...
	18. non worrisome	0	4		24	0		3	0	
	19. outdoor	0	0		0	30		0	0	
	⋮	⋮	⋮		⋮			⋮	⋮	
⋮		⋮	⋮							

《Question list》

Image by values   name [ ]

Car Name	Language of Image
CELSIOR	multiple answers
CROWN	
⋮	

vehicle names and body styles to explore vehicle values by means of images.[18]

A questionnaire card written with 45 domestic and foreign vehicle names, including Toyota vehicles, is given to each panelist to make him or her state the reasons in terms of its image (multiple answers allowable) why he or she would select each vehicle as its imaginary owner. Table 1 shows the result. Vehicle names are listed horizontally with the language of images listed vertically. The figures in the table are the numbers of answers. The 200 images answered are classified into groups of similar ones by drawing a diagram of affinity.

In case of encircled Corona (Toyota vehicle), for example, many answered within ones means and modest in the status group. In the life-style group. on the other hand, familism and nonworrisome are answered frequently. To collect those similar in the sense of value from this table, a cluster analysis is performed as shown in Fig. 6. The sense of value can be classified into nine groups, (1) to (9),

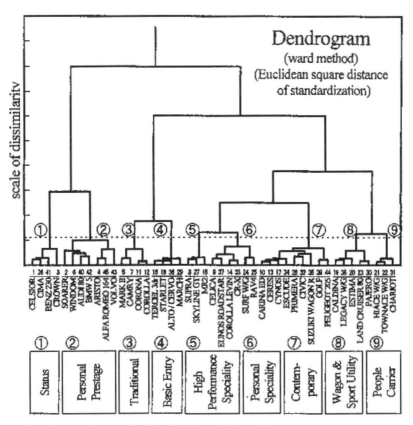

Fig. 6. Grouping to find similarities of the vehicles by cluster analysis

as indicated by arrow lines. The leftmost group of Celsior, Cima, Benz and Crown is expressed by the images of an industrialist and successful man, and cab be named as "status" by the intuition of designers. The adjacent group of Soarer and BMW 518i is expressed as "personal prestige, and the succeeding groups as shown in the figure. This classification result is reasonable and agrees with designers' senses of values.

Next to explore grouping of vehicles by senses of values, a principal component analysis is carried out for positioning analysis of vehicle images and senses of values. For example, Fig. 7 shows positioning of principal component marks as grasped in terms of the first and second principal components.

Interpretation of principal component axes according to the calculation results of eigenvalues and eigenvectors of the correlation matrix are as follows. Of the first principal component axis (horizontal axis) having higher contribution, "within one's means" and "expression of oneself" are contrasted on the left and right sides as clearly indicated in the figure. On the second principal component axis (vertical axis), "status" and "selfish" are contrasted on the upper and lower sides to indicate the senses of values. When combined with the result of cluster analysis, 45 models are clustered as encircled in Fig. 7. The senses of values can be named by (1) to (9) in the figure, which agree well with senses of values of designers.

As a result of this analytical approach, useful information such as input of new model vehicles in nonpositioned areas and input of model changed vehicles

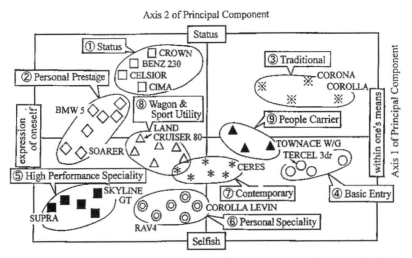

Fig. 7. Positioning analysis of the vehicle image and value by principal
component analysis

having new values in positioned areas for the future design strategy can be component analysis obtained. Through this analytical study approach, it has been confirmed that designers should take the characteristics in each country into consideration in planning the customer-oriented design strategy.

In this study, as shown in Fig. 1, SQC is scientifically applied as customer science to reflect customer feeling in the planning and design stage, so as to make implicit knowledge concerning job quality improvement in the business process in the proactive engineering field into explicit knowledge. This approach method was also applied to "analysis of latent structure of good patents considered by engineers" detailed in references [19, 20] and a similar result has been obtained. Multivariate analysis has also been fixed in the job flow as a useful core technical method not only for qualitative data, but also for enhancement of communication between men and information by making subjective information such as language information into objective information.

4. Conclusion

We have so far proposed a new SQC concept and procedure in the manufacturing industry through introduction of the technical method for practicing scientific SQC. Furthermore, we have verified that mountain-climbing for problem-solving can be executed efficiently using multivariate analysis as the core, in combination with N7 and design of experiment, for solving reactive technical problems and proactive engineering tasks through introduction of two study examples.

References

[1] K. Amasaka, (1996), Application of classification and related methods to SQC renaissance in Toyota Motor, *Data Science Classification and Related Methods*, 684-695, *Springer*.
[2] M. Kamio, and K. Amasaka, (1992), Collection of activity example using SQC method to improve engineering technologies, (in Japanese), *Japanese Standards Association, NAGOYA QC Research Group*.
[3] K. Amasaka, et al., (1996), A study of estimating coefficients of lift at vehicle,

- Using neural networks and multivariate analysis method together -, (in Japanese), *Journal of the Institute of Systems Control and Information Engineers*, 9 (5), 229-237.

[4] K. Amasaka, and H. Sakai, (1996), Improving the reliability of body assembly line equipment, *International Journal of Reliability, Quality and Safety Engineering*, 3 (1), 11-24.

[5] K. Amasaka, and K. Maki, (1991), Application of Multivariate Analysis for the Attraction of Manufacturing Vehicles,(in Japanese), *The Behavior Metric Society of Japan, The 19th Annual Conference*, 190-195.

[6] K. Amasaka, and T. Kosugi, (1991), Application and Effects of Multivariate Analysis in TOYOTA,(in Japanese), *The Behavior Metric Society of Japan, The 19th Annual Conference*, 178-183.

[7] K. Amasaka, and K. Maki, (1992), Application of SQC Analysis Soft in Toyota,(in Japanese), *QUALITY, Journal of the Japanese Society for Quality Control*, 22(2), 79-85.

[8] K. Amasaka, et al., (1995), Aiming at Statistical Package using in the Job Process,(in Japanese), *Journal of the Japanese Society for Quality Control, The 25th Annual Technical Conference,* 3-6.

[9] K. Amasaka, et al., (1996), Influence of Multicollinearity and Proposal of New Method of Variable Selection, -A Study of Applied Multiple Regression Analysis for Analysis of Source of Valuation-,(in Japanese), *Japan Industrial Management Association*, 46(6), 573-584.

[10] K. Amasaka, et al., (1996), A Study on Validity of the BN method for Variable Selection, - A Study of Applied Multiple Regression Analysis for Analyzing Source of Variation Factors (Part II) -, (in Japanese), *Japan Industrial Management Association*, 47(4), 249-256.

[11] K. Amasaka, (1993), SQC Development and Effects at TOYOTA,(in Japanese), *QUALITY, Journal of the Japanese Society for Quality Control, 23(4), 47-58.*

[12] K. Amasaka, (1995), A Construction of SQC Intelligence System for Quick Registration and Retrieval Library, - A Visualized SQC Report for Technical Wealth, *Lecture Notes in Economics and Mathematical Systems*, 445, 318-336, *Springer.*

[13] K. Amasaka, et al., (1992), A Construction of SQC Information Synthetic Network for Accumulating Technical Wealth,(in Japanese), *Journal of the Japanese Society for Quality Control, The 22nd Technical Conference*, 37-40.

[14] K. Amasaka, et al., (1996), The Promotion of Science SQC in Toyota,

Proceeding on The International Conference on Quality, 565-570, *Yokohama Japan.*

[15] K. Amasaka, et al., (1994), Consideration of effieientical counter measure method for Foundry,(in Japanese), *Journal of the Japanese Society for Quality Control, The 47th Technical Conference*, 60-65.

[16] K. Amasaka, et al., (1995), The Q.A. Network Activity for Prevent Rusting of Vehicle by Using SQC,(in Japanese), *Journal of the Japanese Society for Quality Control, The 50th Technical Conference*, 35-38.

[17] K. Amasaka, et al., (1996),A Improvement of Disk Brake Quality, (in Japanese), *Journal of the Japanese Society for Quality Control, The 53rd Technical Conference*, 89-92.

[18] K. Amasaka, et al., (1996), A Study of Effectiveness of SQC for Management,(in Japanese), *Journal of the Japanese Society for Quality Control, The 53th Technical Conference*, 85-88.

[19] K. Amasaka, and M. Ihara, (1995), Latent Structure of Goodness of Invention, *Lecture Notes in Economics and Mathematical Systems*, 445, 348-353, *Springer*.

[20] K. Amasaka, et al., (1996), Latent Structure of Engineers Attitudes to the Goodness of Invention and Patent (No.1) and (No.2), (in Japanese), *Journal of the Japanese Society for Quality Control, The 52nd Technical Conference*, 17-24.

Chapter 3: "Science SQC" Proposal

Utilizing Management SQC:

A Study on "Science SQC" for Improving the Job Quality

Looking at changes in the environment surrounding the manufacturing industry, corporate efforts to foster the development capabilities of engineers for new products are again being called for. This study outlines "Science SQC" by utilizing "Management SQC" as a new SQC Concept and Procedure in the manufacturing industry for improving the job quality of engineers and managers in each stage of the manufacturing business and process with excellent QCD, and introduces application examples at Toyota.

Keywords: "Science SQC", Improving the Job Quality, "Management SQC", "Scientific SQC", "SQC Technical Methods", and manufacturing industry.

1. Introduction

Recently a review of the changes in the environment surrounding the manufacturing industry indicates an ever greater necessity for corporate efforts to amplify and capitalize upon the technical progress of engineering staff who bear the main burden of these times. For this purpose, it is essential to improve their work qualities and engage in business cycles under optimum conditions.

The statistical quality control (SQC) provides a method for analyzing complicated events and cause/result relationships. If engineering staff keep a careful watch over their work observing actual products with care in the workshop and apply SQC properly, SQC can assist them in remedying work processes and effectively raise job quality. The importance of SQC as a behavioral science must be recognized in order to contribute in creating proprietary engineering by utilizing it form a scientific point of view, not only for solving existing problems, but also for potential and future problems.

This report proposes "Science SQC" by utilizing "Management SQC" as a new SQC concept and practice for a manufacturer that can be systematically and organically operated. Further, this report verifies that the method can improve the job quality of engineering staff at every business process and contribute to establishing excellent QCD for manufacturing excellent quality products by showing some actual examples at one manufacturer - the Toyota Motor Corp.

2. Consistency in "SQC Promotion Cycle"

Development of the SQC promotion cycle illustrated in Fig.1 is essential for continuously increasing expectation to and functions of the SQC activity. Amasaka [1] has promoted "SQC Renaissance" (since 1988) at Toyota, one of the manufacturers, and verified its effect. Practically, SQC has been applied to today's engineering themes and results obtained as proprietary and management technologies. Another aspect of the objectives is to develop SQC promotion cycle activities in which SQC practices result in practical and full development of SQC education that facilitates effective development of human resources, which in turn, will be reflected in the performance of operations.

This promotion is the essence of the "Science SQC" which is discussed throughout this report. Fig. 1 schematically illustrates the concept of the "SQC Promotion Cycle" at a manufacturer.

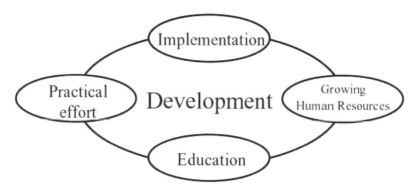

Fig. 1. Schematic drawing of "SQC promotion cycle"

3. Proposal on and Implementation of "Science SQC" [2-4]

The "Science SQC" being proposed here consists of two cores; one is "Scientific SQC" and the other, the "SQC Technical Methods". This scientific method was the starting point for change from "SQC Renaissance" to "Science SQC" (started in 1995).

3. 1. Implementation of "Scientific SQC"

If engineers and management of a manufacturer (herein called businessmen) fully utilize SQC in challenging important technical themes, excellent QCD studies can be achieved by adding insight to the scientific and inductive methods to the routine way of job performance in addition to engineers' deductive work methods. It is important to implement SQC in each business process after determination of a desired picture under consideration of problem construction, theme setting and goal achievement for solving existing technical themes as well as also potential and future themes, so as not to use SQC simply for statistic or trial-and-error analyses.

Based on this, businessmen can improve proprietary and management technologies by using a logical way of thinking and clarifying the gap between logic and actual. Thus, they can create universally applicable answers and

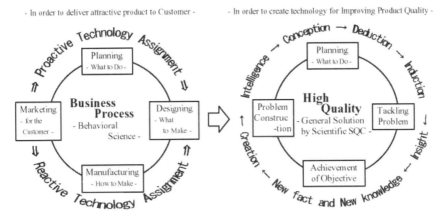

2-1 Business Process for Customer Science   2-2 Scientific SQC for Improving Technology

Fig. 2. New schematic drawing of "Scientific SQC" to conduct superior QCDS research

improve the job quality, without staying with a special or partial answer to an individual problem. As a result, versatile SQC activities contribute to creating properties, which will be passed on to the next activities for further development. Construction of an excellent concept for a new technical theme is permitted through use of the created intelligent and useful information, thereby further improving job quality. Fig. 2 schematically illustrates the concept of "Scientific SQC" which makes excellent QCD study possible, which is the basis of "Science SQC" as a general problem solving method, seeking universally applicable answers.

3. 2. Establishment of the "SQC Technical Methods"

Using proprietary technologies and acquired knowledge, SQC resolves complex entanglements of cause and effect relationships for both quantitative and qualitative data. Hence, it has become a highly convenient method of technological analyses for improving proprietary technology. Through surveys and use of accumulated technologies for solving today's technical themes, N7 and basic SQC methods enable full support for designing experimental and analytical processes, thus making it possible to analyze technology promptly and with error-free thinking.

In addition by capitalizing upon proprietary technology, the use of the multivariate analysis method enables 70% to 80% of the mileage required to go before finding solutions to potential or future problems to be covered. Combined with SQC methods such as design of experiment, the remaining distance can be covered effectively. Fig. 3 schematically illustrates the concept of "SQC technical methods" for conducting "Scientific SQC". Detailed practical reports are shown in references [5-6]. Use of SQC methods effectively brings about expected technical results, using analysis of sources of variation, modeling for prediction and control, and concurrent application with neural networks as the new technical method.

Above all, multivariate analysis, which allows flexible technical analysis of various data collected in the past, has become more widely used not only as an analytical method by SQC specialists but also as a technical method by Toyota engineers in the upstream production engineering department requiring quick development of advanced design technologies, in the production preparation department in the middle reaches requiring efficient development of new

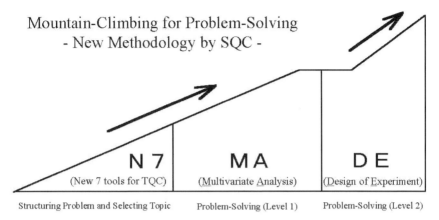

Fig. 3. Schematic drawing of "SQC technical methods"

production engineering and control technologies, and in the downstream manufacturing department requiring stabilized high-quality production for improving production technologies.

Thus, the method is actively being studied as a key technology for a variety of fields, mainly at Toyota Motor Corp. [7]

4. Effectiveness of "Management SQC" [3, 8]

An important aspect of the SQC activity expected by a manufacturer lies in the support for solving deeply rooted technological themes. That is, it is expected to scientifically clarify unknown areas of outstanding chronic and bottleneck problems for improving the technical level, thus, contributing to the development of new technologies, new processes and new materials. For these reasons, further development of "Science SQC" is needed.

In the practical development, it is required to approach universally applicable solutions without staying with an individual solution, by scientifically analyzing disagreements with the principles and fundamental rules as a technological problem and elucidating the six gaps shown in Fig. 4 between the theory, calculation, experiment and actual result. To analyze these gaps is not easy since problem solving is dependent upon each individual's problem analysis capability. In many cases, filling these gaps boils down to an organizational problem. To solve such a problem, it is important to clarify the six gaps between planning, designing, manufacturing and marketing organizations (Fig. 4), that is, unknown factors of business processes must be clarified for better communications between departments.

The manager who handles the business resources in each workshop must clarify business processes among organizations by practicing task management team activities within and among departments including part suppliers in cooperation with the staff, in order to realize common possession and construction of problems and improve the decision quality for problem solving. The organization standpoint development method where "Science SQC" is used for management is called "Management SQC." Fig. 4 schematically illustrates this using a conceptional drawing. Recently, Amasaka and others evaluated the effectiveness of "Management SQC" by discussing practical examples [9] at Toyota Motor Corp.

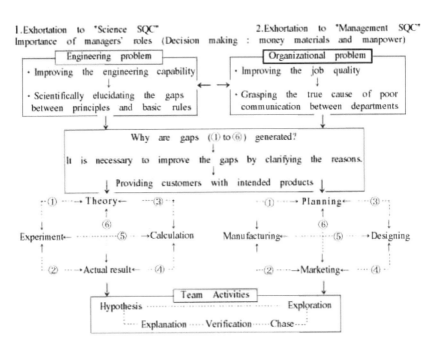

Fig. 4. "Management SQC" for "Science SQC" development

5. Application Example

Analysis of Disk Brake Pad Squeal Noise Factors [9]

5. 1. Objective

The squeal noise of a disk brake occurs when the brake vibrates due to delicately unstable contact between the brake pad and the disc rotor (herein called the pad and rotor) on braking. This is a never-ending problem for the automotive industry worldwide. [10] The objective is to develop "Science SQC" by systematically utilizing "Management SQC" for analyzing the factors and improving production quality.

5. 2. Task Management Team Activity

Mixing 10 - 20 types of materials and sintering the mixture produces the pad, which is one of the main parts of the brake. Since compatibility with the suspension, rotor and caliper is also required, development and design departments of the automobile and brake makers must perform jobs jointly under optimum conditions. For this reason, to construct a task management team is a useful method for improving the technical levels of both parties, hence know-how can be shared by both parties.

The target of "Management SQC" is that both parties treat design and manufacturing problems as common technical problems by combining their target consciousness into one, not by treating them as partial problems to be solved individually. This permits optimization of independent processes as a total business process.

5. 3. Use of the "SQC Technical Methods"

The three steps (market research, cause and result relationships analysis and process improvement) shown in Fig. 5 are sequentially developed for full utilization of the "SQC Technical Methods" in order to positively and scientifically clarify design and manufacturing problems in order to evaluate the

"Science SQC" effect. In practice, a program is implemented by synchronizing the technical management, which manages the design technology, and the product management which manages the manufacturing technology.

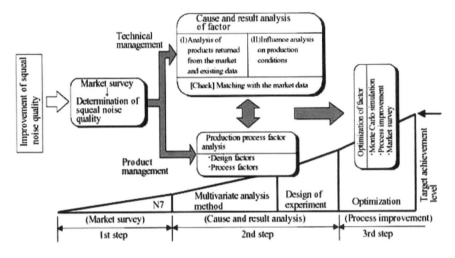

Fig. 5. "Management SQC" activities process

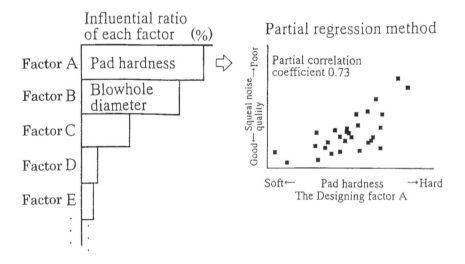

Fig. 6. Influential effect of each factor

5. 4. Factor Analysis

5. 4. 1. Use of SQC in technical management

In factor analysis (I), the squeal noise level is measured using a volume indicator to obtain a set of data; one being the squeal index which is the objective variable, and the other, the pad's physical characteristics which are the explanatory variable. Data are collected from the market and previously accumulated. As a result of squeal noise factor analysis using the multivariate analysis method, it is known that the factor relative to the pad has the largest influence among all explanatory factors evaluated. The result agrees with the experience technology.

At the next stage of factor analysis (II), a spot beam is directed onto the pad and an experiment using an orthogonal table is planned and implemented, aiming at optimum parameter and tolerance design, including new physical characteristic factors and expanding the level range. Based on the analysis results of the dispersion and multiple regression analyses shown in Fig. 6, it is determined that design factor A (pad hardness) and B (pad blowhole diameter) are highly influential. In order to allow optimum design of each design factor, the optimum level of product quality is established using the partial regression plot method. For example, the effect of pad hardness shown in the figure used to be known qualitatively only, which is now clarified quantitatively.

It is newly gained knowledge that the effect of the rough density (blowhole diameter) expresses the area density of the pad's internal organization after sintering. Effects of other design factors ($C, D, E,$ etc.) are clarified in the same manner. Technical matching of analysis results and market evaluation is confirmed and verified using monitor vehicles and by bench tests.

5. 4. 2. Use of SQC in product management

Next, cause and result analyses are conducted on relationships between the squeal noise and fluctuating variables in the manufacturing processes, using similar methods. Optimum process control conditions for optimum production are determined by quantitatively evaluating the oven temperature during pad sintering (production factor F), temperature and humidity at pad forming (production factor G) and other factors.

Eq. (1), which can be obtained from the multiple regression analysis results, is the estimated model equation for quality improvement and is derived from a positive scientific standpoint.

[Squeal noise index]$=a_1$ (Design factor A)$+a_2$ (Design factor B)$+ \cdots a_6$
(Production factor F) $+ a_7$ (Production factor G) $+\cdots+b$,   (1)
Where, a is partial regression coefficient and b is a constant.

5. 5. Factor optimization and process improvement

In order to determine the optimum conditions for all factors being evaluated, Monte Carlo simulation is carried out to obtain distribution of the squeal noise quality (index) using equation (1). Fig. 7 illustrates the forecast effect of changing design factor A (old to new pad hardness).
When the process improvement effect is forecasted by optimizing the combination of major design and production process factors, it is judged that market claims can be reduced to 1/3 - 1/5. As a result of the market survey after process improvement, market claims were reduced to 1/4, which proves that the forecast and actual result agree.

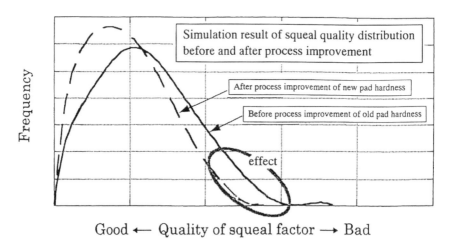

Fig. 7. The forcast effect of changing dsign factor A (old to new pad hardness)

5. 6. Result

The technical levels of both parties are improved through the task management team activities conducted in cooperation with the automotive and brake makers. Thus, the effectiveness of "Science SQC" using "Management SQC" is verified.

6. Conclusion

We have proved the effectiveness of "Science SQC" by utilizing "Management SQC" being proposed as a new concept and procedure for implementing SQC by manufacturers based on the successful practice at Toyota Motor Corp. In the future, we wish to accumulate positive research on the application of "Science SQC" to a wide range of business themes in order to establish a next-generation scheme, which will be a universal SQC method.

References

[1] K. Amasaka, (1993), SQC Development and effects at Toyota, (in Japanese), Quality , 23 (4), 47-58.

[2] K. Amasaka et al., (1996), The promotion of science SQC in Toyota, *Proceedings of the International Conference on Quality*, 565-570.

[3] K. Amasaka, (1998), Application of classification and related methods to SQC renaissance in Toyota Motor, *Data Science, Classification and related methods,* 684-695, *Springer.*

[4] K. Amasaka, (2000), A demonstrative study of a new SQC concept and procedure in the manufacturing industry - Establishment of a new technical method for conducting SQC - , *An International journal of mathematical & computer modeling*, 31 (10-12), 1-10.

[5] T. Takaoka, K. Amasaka, (1991), Derivation of statistical equation for fuel consumption in S. I. engines, (in Japanese), *Quality*, 21 (1), 64-69.

[6] K. Amasaka et al., (1993), A Study of Quality Assurance to Protect Plating Pants from Corrosion by SQC, (in Japanese), *Quality,* 23 (2), 90-98.

[7] K. Amasaka, T. Kosugi, (1991), Application and effects of multivariate analysis at Toyota, (in Japanese), *The Behavior Metric of Japan, The 19th Annual*

Conference, 178-183.

[8] K. Amasaka et al., (1996), A Study of effectiveness of SQC for Management, (in Japanese), The 53rd Technical Conference, *Journal of the Japanese Society for Quality Control*, 85-88.

[9] K. Amasaka et al., (1996), A study on improving disk brake pad quality to reduce squeal, (in Japanese), *The 53rd Technical Conference, Journal of the Japanese Society for Quality Control*, 89-92.

[10] N. Miller, (1978), An analysis of disc brake squeal, *SAE Technical Paper* no. 780332.

Chapter 4: "Science SQC" Implementation

New SQC Internal Education

A Proposal of "Science SQC" for Improving Principle of TQM

Recently, it has been of important to challenge the manufacturing production demands of the 21st century. This report proposes a new SQC concept, "Science SQC" as a demonstrative-scientific method, which enables us to improve the principle of TQM improve systematically. The objective of this approach is to indicate the application of a New SQC Internal Education through implementing "Science SQC", which leads to promoting organically the business process quality. The performance of the proposed method and the results are given through studies and practices at a one manufacturing - the Toyota Motor Corp.

Keywords: New SQC Internal Education, Science SQC, total business process quality, Inprove the Principle of TQM, TQM Activities at Toyota, the TPS, TDS and TMS.

1. Introduction

One expected role of SQC (Statistical Quality Control) in the manufacturing industry is to assist in solving foreseen latent engineering problems in addition to existing ones. In other words, development of the new SQC as a scientific methodology is required to clarify scientifically implicit technical knowledge concerning pending chronic and bottleneck engineering problems so as to contribute to the development of new technologies, processing methods and materials. Recently, it has been of importance to challenge of the manufacturing production demands of 21 st century.

This rcport proposes a new SQC concept "Science SQC" as a demonstrative-scientific method, which enables us to improve the principle of total quality management TQM systematically. The objective of this approach is to indicate the application of a new SQC internal education through implementing "Science SQC", which leads one to promote organically the business process quality. The performance and the results are given through studies and practices at a one manufacturing - the Toyota Motor Corp.

2. Need for new SQC to Improve the Principle of TQM [1, 2]

2.1. TQM Activities in Toyota

TQM activities for working under optimum conditions so that problems can be eliminated by improving job quality are required in order to manufacture attractive products that satisfy customer needs. Fig. 1 illustrates the circle showing the relationships between the TMS, TDS and TPS, which are main concepts of Toyota activities and management technologies. In other words, all departments from planning, developing and design to production and sales should grasp the fundamentals of manufacturing at each stage of the business process and establish a new methodology for improving the level of engineering through a demonstrative study.

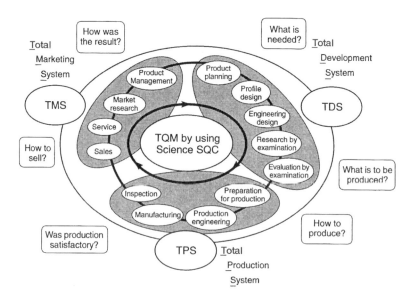

Fig. 1. TQM activities at Toyota

2.2. Need for New SQC

It is well recognized that the SQC has played an important role in the wide adaptation of TQM, when the quality control development history is reviewed. However, a new SQC is necessary under the rapid innovation of technology, changes in the corporate environment and requirements for total production shown in Fig. 1. Aiming for further improvement of the total business process quality, it is the today's theme for the corporate to incorporate the new SQC for a challenging new theme.

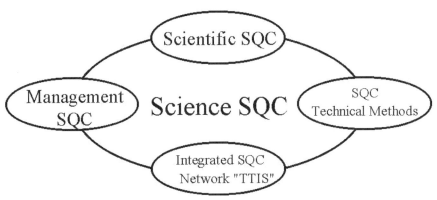

Fig. 2. Schematic drawing of "Science SQC"

3. "Science SQC" Proposal and Demonstration [2-5]

A practical customer science for converting customers' wishes into engineering terms is necessary for conducting demonstrative studies today. It is important that all departments have the same objective awareness of unified cooperative activities in order to clarify implicit knowledge in the business process. As a scientific methodology for this purpose, systematic and organizational SQC application based on a new concept designed to link organically the four cores shown in Fig. 2 is called "Science SQC".

3.1. Practical Application of "Scientific SQC"

The primary objective of SQC in the manufacturing industry is to enable all engineers and mangers (hereafter called businessmen) to attain excellent R&D activities related to quality, cost and delivery (QCD) through insight obtained by applying SQC to scientific and inductive approaches in addition to the conventional deductive method in tackling important engineering problems. As shown in Fig. 3, it is important to depart from mere SQC application for statistical analyses or trial-and-error type analysis but to use SQC scientifically in each stage from problem structuring until goal attainment by grasping the desirable form.

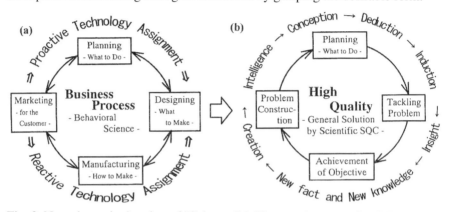

Fig. 3. New shematic drawing of "Science SQC" to conduct superior QCD
 research. (a) Business process for customer science. (b) Scientific SQC for
 improving technology.

3.2. Use of "SQC Technical Methods"

For solving today's engineering problems, it is possible to improve the experiment and analysis plans by using N7 (New Seven Tools) and basic SQC method based on the investigation of accumulated technologies. Furthermore, it is possible to reach the crest of problem solving mountain by proactive combination of the design of experiments as required with multivariate analysis amalgamated with the engineering technology. As shown in Fig. 4, the methodology in which the SQC method is used in combination at each stage in problem solution has spread and rooted as "SQC Technical Methods" for efficient improvement of the jobs of businessmen.

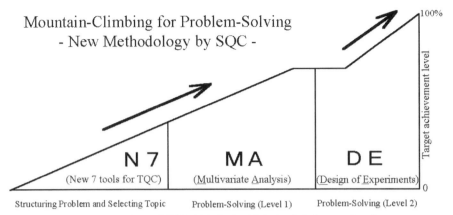

Fig. 4. Schematic drawing of "SQC Technical Methods" for conducting "Scientific SQC"

3.3. Construction of Integrated SQC Network "TTIS"

Various successful SQC applications in actual business have contributed to forming engineering assets, but systematization of SQC applications is necessary for inheritance and further development. It is a prerequisite for the development of "Science SQC". This requirement is satisfied by the "TTIS" (Total SQC Technical Intelligence System) for SQC information synthesis networks which supports solving engineering problems. As shown in Fig. 5, the "TTIS" is an intelligent system for SQC applications which consists of four main systems synthesized to grow while supplementing one another.

3.4. Practical Application of "Management SQC"

The main objective of SQC in the manufacturing industry is to support quick solutions to deep-rooted engineering problems. As shown in Fig. 6, especially in the application of "Science SQC", the differences from the principles and rules in an engineering problem should be analyzed scientifically to clarify six gaps between the theory, calculation, experiment and actual result to obtain a generalizable solution. Filling up these gaps results is an organizational problem.

For problem solution, it is necessary to clarify six gaps between the planning, designing, manufacturing and marketing departments or to clarify the implicit knowledge on the business process for good coordination among departments.

The methodology for organizationally managing the development of "Science SQC" is called "Management SQC". In recent years, Amasaka et. al.[3-5] have been discussing the effectiveness of "Management SQC" at Toyota Motor Corp.

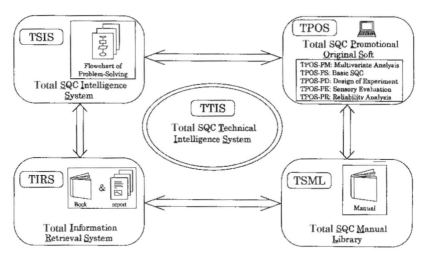

Fig. 5. A schematic drawing of "TTIS" for SQC information sysnthesis networks

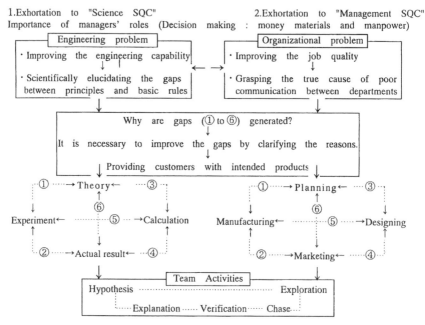

Fig. 6. "Management SQC" for "Science SQC" development

4. Consistency in "SQC Promotion Cycle" [6]

Development of the "SQC Promotion Cycle" illustrated in Fig. 7 is essential for continuously increasing expectation and functions of the new SQC activity. Practically, SQC has been applied to today's engineering themes and results obtained as proprietary and management technologies. Another aspect of the objectives is to develop SQC promotion cycle activities in which SQC practices result in practical and full development of SQC internal education that facilitates effective development of human resources, which in turn, will be reflected in the performance of operations. This promotion is the essence of the "Science SQC" which is discussed throughout this report. Fig. 7 illustrates schematically the concept of "SQC Promotion Cycle" at a manufacturer[1].

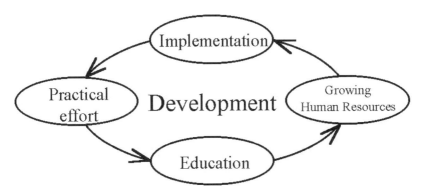

Fig. 7. Schematic drawing of "SQC promotion cycle"

5. The Promotion of New SQC Internal Education [7]

5.1. Need for New Internal SQC Education

SQC internal education in the company should be higher in level than mere education for studying the SQC. The true value of education is evaluated by to what extent the SQC is reflected in actual results. Conventional internal seminars have mainly covered manual calculations, consisting mostly of a few existing curricula, with emphasis laid on analysis. Recently, it is considered that the SQC personal computer software must be utilized, giving importance to the practical design and analytical design, so that what is learnt will be useful for actual jobs, keeping pace with the rapidly advancing technology. For above reasons, such methods should be combined for utilizing "SQC technical methods" along the flow of jobs by starting from N7 to multivariate analysis and experiment planning method, for achieving the scientific approach.

5.2. Companywide Promotion of New SQC Internal Education

In order to develop "SQC Promotion Cycle" shown in Fig. 8, establishment of the organization and system is essential for realizing company-wide development of the new SQC internal education systematically. One of the important subjects is to establish systematic SQC internal education systems for beginners to upper class levels by clarifying the objectives so that the SQC will be useful for a variety of jobs assigned to departments from the upper to down streams.

The second important point is implementation of SQC internal education must have sufficient quality and quantity to match today's needs. To achieve this target, it is essential to hold excellent seminars with good trainers, good textbooks and good software, as shown in Fig. 8. In practice, self-prepared plans giving importance to actual job processes and their operation, planning of curriculums useful for actual jobs and operation of seminars by businessmen are essential.

The third important point is that personnel who have complete the SQC seminar will have improved problem solution capabilities, and trained workers actually challenge problems at individual workshops to demonstrate actual results. By reflecting the results thus obtained to the SQC internal education, "SQC Promotion Cycle" can be developed further.

6. Example of SQC Internal Education [1, 8]

6.1. Arrangement and Fulfillment of Hierarchical SQC Seminar

We used to train the SQC basic mainly by calculation methods, reliability analysis and design of experiment. However, they were not sufficient for achieving development of the "SQC promotion cycle" illustrated in Fig. 8. From the standpoints of Sections 2-5, the hierarchical SQC internal education and

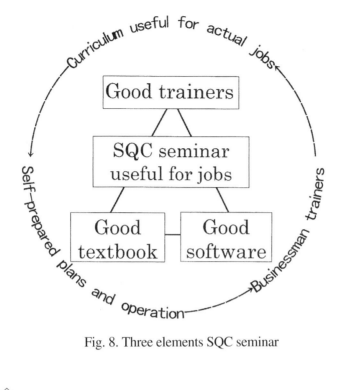

Fig. 8. Three elements SQC seminar

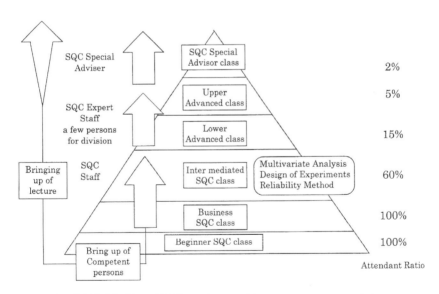

Fig. 9. Divided into SQC seminar and human resources training

system were planned and implemented in six ranks: beginner, business, middle class, upper class, highest class and advisor as shown in Fig. 9. Twelve thousand businessmen were allocated to each ranking of the six stage step-up course at the ratios as shown, under full consideration of improving human resources.

It is requested that all personnel who are allocated to the Beginner and Business classes attend respective courses, which are sufficient to cover the daily work, while at the middle class course, participants will learn and practice the new SQC methods. In the upper rank course and higher, the systems are established aiming at trainers and leaders of respective workshops. Qualifications for SQC special staff and SQC special advisor are determined and the respective titles are given to successful candidates. Eight hundred special staffsers and advisors who successfully completed the six steps are actively engaged in SQC seminars as trainers and as SQC promotion leaders of workshops of 200 departments.

6.2. Curriculums and Contents of New SQC Seminar

In the case of the hierarchical SQC seminar, three courses are prepared for the Beginner-ranked personnel: technician, sales and clerical courses (100 participants per course, one seminar per year or more). Fig. 10 shows typical curricula of the technician course (3 days per course, 21 hours per course). The aforementioned "SQC technical methods" are used and the course contains 12 lectures (SQC Review: SR 1 to SR 12).

In order to provide friendly seminar useful for actual jobs by businessmen, the SQC personal computer software TPOS (Toyota TQM Promotional SQC Original Software) is used at SR 2 . As you can see from Fig. 11, trainees operate the TPOS according to the solution procedure of an actual problem-solving activity using the SQC methods, to understand the attractive "Science SQC" and the usefulness by exercise.

For application to actual jobs, it is planned to make trainees learn that to combining various SQC methods is important for obtaining successful results. Then, trainees will learn the necessary theory and practice of SQC methods at SR 3 to SR 9. At these lectures, the TPOS is fully utilized, followed by SR 10 examination. Then, SR 11 and SR 12 are prepared, where practical use of "Science SQC" is studied according to purposes of each course, followed by question and answers and discussion among participants to complete the course. It is planned that each trainee will bring the TPOS back to his/her own workshop for practical use.

73

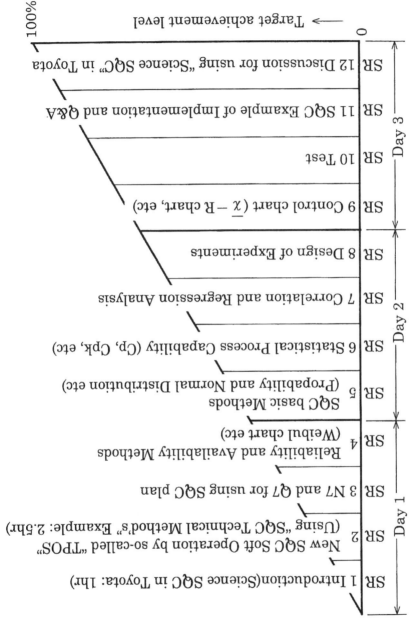

Fig. 10. Curriculum of SQC's beginner's class

SR 1 Introduction(Science SQC) in Toyota: 1hr)

SR 2 New SQC Soft Operation by so-called "TPOS"
(Using "SQC Technical Method's" Example: 2.5hr)

SR 3 N7 and Q7 for using SQC plan

SR 4 Reliability and Availability Methods
(Weibul chart etc)

SR 5 SQC basic Methods
(Propability and Normal Distribution etc)

SR 6 Statistical Process Capability (Cp, Cpk, etc)

SR 7 Correlation and Regression Analysis

SR 8 Design of Experiments

SR 9 Control chart ($\bar{X}$ – R chart, etc)

SR 10 Test

SR 11 SQC Example of Implementation and Q&A

SR 12 Discussion for using "Science SQC" in Toyota

← → Target achievement level

100%

0

Day 1

Day 2

Day 3

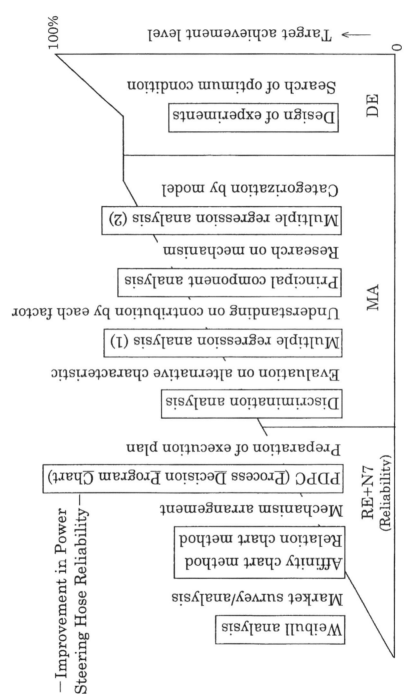

Fig. 11. Example of mountain-climbing for problem-solving in PC exercise (full utilization of "SQC technical methods")

7. "Science SQC" Application Example [9]

The "brake pad quality assurance activity" is adopted as a typical recent study activity for systematic development of the concept and methods of "Science SQC". This is a typical example where engineers, who completed the new SQC training courses explained in Sections 5 and 6 challenged under instruction by the SQC special staff and advisors.

7.1. Objective

In the disc brake, the caliper pushes the pad against the rotor for braking. The brake squeal or abnormal noise is generated when the contact between the pad and rotor becomes subtly unstable. The likeliness of generating squeal or abnormal noise and the braking performance are contradictory phenomena. The main objective of this activity is to establish a technology for attaining satisfactory braking performance while minimizing squeal through exchange of the result of sensitivity analysis on the causes of squeal and braking performance in the design engineering field and the know-how in the production engineering field.

7.2. "Total Task Management Team Activities"

Forming an organization allowing use of know-how and information in among the vehicle design, parts design, production process design, manufacturing, inspection, maintenance and marketing (service and market quality) departments for vehicle production including the assembly manufacturer will lead to improvement of engineering capability through supporting conception by engineers, possibly resulting in improvement of the business process quality.

This organization has been realized in the form of a "Total Task Management Team" as shown in Fig. 12. Five teams, QA1 (engineering design), QA2 (production engineering), QA3 (manufacturing and inspection), QA4 (equipment maintenance) and QA5 (quality and production information), are linked mutually for total QA network (QAT) activities.

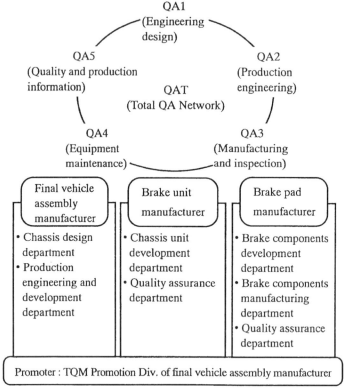

Fig. 12. Organizational outline of "total task management team"

7.3. Development of "QAT"

As the key technology for quality assurance, three management activities "total technical management (TM)", "total product management (PM)" and "total information management (IM)" were proposed to promote "QAT" [1, 5] were promoted by integrating these activities as shown in Fig. 13. In actual practice, "TTIS" for SQC information synthesis was used with full utilization of [3, 4] in designing the experiment and analysis so as to avoid a trial-and-error process.

(1) "Total Technical Management"
This activity focuses on reviewing the work method and control items in process conditions by clarifying how brake performance and squeal are affected by dispersions in raw material properties and process conditions through sensitivity analyses.

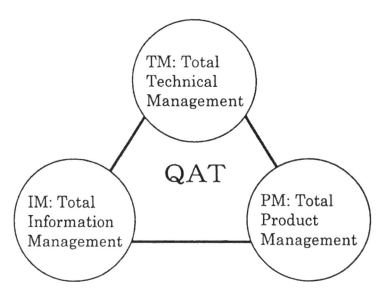

Fig. 13. Three management activities for "total QA network"

(2) "Total Product Management"

This is a production activity for realizing the process conditions and control items using the production QA network table based on the matrix chart and process FMEA methods, in which the design, production engineering, manufacturing, maintenance and quality assurance departments participate jointly.

(3) "Total Information Management"

This activity focuses on establishing the system for timely feedback of the market quality information (from dealers), next process information (from vehicle assembly manufacturer) and local process information (parts manufacturer) to respective processes.

7.4. Factor Analysis

For quick systematic and organizational optimization of the total QA network shown in Fig. 13, the "SQC technical methods" as shown in Fig. 14 is used. Each type of arrow in the figure represents the activity of each team among QA1 to QA5.

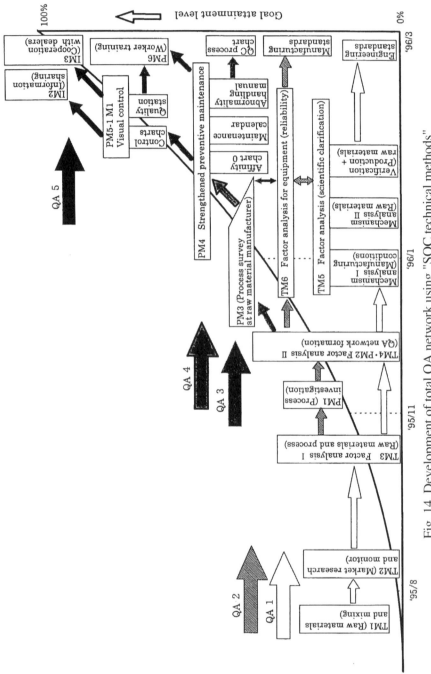

Fig. 14. Development of total QA network using "SQC technical methods"

(1) "Total Technical Management"

Analyses of each raw material (TM1 in Fig. 14) and market research (TM2 in Fig. 14) were conducted, and factor analysis based on the results (sensitivity analysis) was performed to screen the materials in a short period. In factor analysis I (TM3 in Fig. 14) using principal component analysis, for example, it was found that the embedded mineral grain size and the inorganic fiber diameter are related to the abnormal sound characteristic and wear, respectively as shown in Fig. 15. Each a represents 'a' region where abnormal sound or wear is very conspicuous, while each 'b' represents a region where its influence remains. Each 'c' has been found as a region where both characteristics are not contradictory.

Factor analysis I has shown that variation in the production process conditions (PM1 in Fig. 14). For the abrasive components (inorganic fibers and hard particles), technical analysis such as electron microscopy to observe the states of dispersion in each raw material and the pad was used in combination to verify the factorial effect. As a result, the mechanism of braking performance variation due to dispersion in inorganic fiber production has been clarified to enable improvement in cooperation with the material manufacturer.

In the production preparation stage, product drawings were drafted using the technical QA network based on the market quality requirement (TM4 in Fig. 14) and equipment drawings were drafted using the production QA network based on the production quality requirement (PM2 in Fig. 14). Important factors related to raw material acceptance, production process conditions and state management

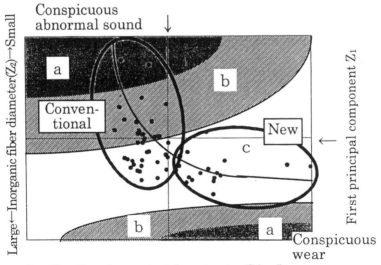

Fig. 15. Example of analyzed influences of raw material properties

were sorted out in this stage.

For the sorted out factors, the phenomena were analyzed (TM5 in Fig. 14) to determine the equipment conditions quantitatively and the product drawing tolerances were determined by combined use of the design of experiments and multivariate analysis for scientific optimization. For example, use of the partial residual regression plot has clarified that the thermoforming temperature has a causal relation with the alternative characteristic of squeal as shown in Fig. 16.

Each 'a' represents a region where the strength is insufficient or a region not suitable for forming while each 'b' represents a region where the influence remains. In consideration of the strength and formability, region 'c' has been determined as the control conditions of the thermoforming process. To ensure control conditions shown by region 'c', factor analysis on the equipment side (TM6 in Fig. 14) for the molding die temperature dispersion, for example, was conducted, thus enabling unification of the pad forming temperature.

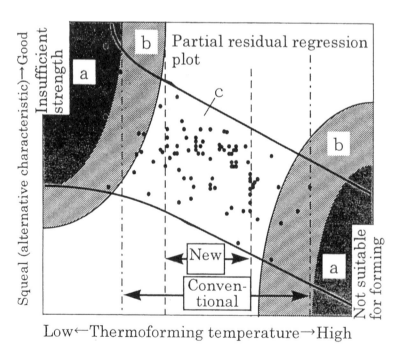

Fig. 16. Casual relation between alternative characteristic of squeal and thermoforming temperature.

(2) "Total Product Management"

To attain the process conditions clarified by total technical management, the defect causes were studied by process investigation (PM1 in Fig. 15) and the relationships with the quality characteristic were checked by the production QA network (PM2 in Fig. 14). Further to complete the QA network, the brake unit manufacturer, brake pad manufacturer and raw material manufacturers formed a task team for mutual quality reviewing (PM3 in Fig. 14).

As a result, the current process capability was clarified with summarized recurrence and release prevention measures, leading to strengthened preventive maintenance (PM4: maintenance calendar in Fig. 8), visual control (PM5: inline SQC in Fig. 14) and worker training (PM6: abnormal quality handling manual in Fig. 14).

(3) "Total Information Management"

Quality check station (IM1 in Fig. 14) was formed to enable the process information as the result of total product management to be seen immediately. The route for making the market information (DAS: Dynamic Assurance System for high Reliability and Quality [17]) owned by the final vehicle assembly manufacturer available to parts manufacturers (IM2 in Fig. 14) and the route for abtaining the actual product information from dealers (IM3 in Fig. 14) were formed.

7.5. Effect

Through practical use of "Management SQC" by the total task management team, the business process quality has been improved.

Estimated claim ratio reduction: to 1/4 (from 2.6% to 0.6%)
In-process fraction defective: down 40% (from 0.5% to 0.3%)
Short convergence of initial failures: (from 9 months to 3 months)
Cost reduction: 6.3% (104 Yen/unit)

8. SUMMARY

The objective of this approach is to indicate the application of a new SQC internal education through implementing "Science SQC" which leads to promote organically the business process quality. The performance of the proposed method and the results are given through studies and practices at a manufacturing company, the Toyota Motor Corp, as shown in [11,14].

References

[1] AMASAKA, K and KOSUGI, T., (1997), A proposal of the new SQC internal education for management, (in Japanese), *Journal of the Japanese Society for Quality Control, The 27th annual technical conference,* 19-22.

[2] AMASAKA, K., KOSUGI, T., and YAMAMOTO, M., (1997), The development of new SQC for improving the principle of TQM, (in Japanese), *Journal of the Japanese Society for Quality Control, The 55th technical conference,* 13-16.

[3] AMASAKA, K., (2000), A demonstrative study of a new SQC concept and procedure in the manufacturing industry -Establishment of a new technical methods for conducting scientific SQC-, *An International Journalist Mathematical computer modeling,* 31(10-12), 1-10.

[4] AMASAKA, K., KOSUGI, T., and OHASHI, T., (1996), The promotion of "Science SQC" in Toyota, *Proceeding of the international conference on quality, Japan,* 565-570.

[5] AMASAKA, K., (1997), A study on "Science SQC" by utilizing "Management SQC", *Proceedings of the 14th international conference on production research, Japan,* 730-733.

[6] AMASAKA, K., (1993), SQC development and effects at Toyota, (in Japanese), *Quality, Journal of the Japanese Society for Quality Control,* 23(4), 47-58.

[7] AMASAKA, K. and AZUMA, H., (1991), The practice SQC education at TOYOTA, (in Japanese), *Quality, Journal of the Japanese Society for Quality Control,* 21(1), 18-25.

[8] AMASAKA, K and MAKI, K., (1992), Application of SQC analysis soft in Toyota, (in Japanese), *Quality, Journal of the Japanese Society for Quality Control,* 22(2), 79-85.

[9] AMASAKA, K., (1997), The development of new SQC for improving the principle of TQM in Toyota, *Proceedings of quality symposium, Tainan, Taiwan,* 429-434.

[10] SASAKI, S., (1972), Collection and analysis of reliability information in automotive industries, *the 2nd reliability and maintainability symposium, Union of Japanese Scientists and Engineers,* 385-405.

[11] AMASAKA, K., NITTA, S., and KONDO, K., (1996), An investigation of engineer's recognition and feelings about good patents by new SQC methods, (in Japanese), *Journal of the Japanese Society for Quality Control, The 25th technical conference,* 17-24.

[12] AMASAKA, K., NAGAYA, A., and SHIBATA, W., (1999), Studies on design SQC with the application of Science SQC - Inproving of business method for automotive profile design, *Japanese Journal of Sensory Evaluation,* 3(1), 21-29.

[13] SUGIMOTO, Y., HAYASHI., T, KOIDE, I., and AMASAKA, K., (1997), The development of working conditions taking the lead epoch, (in Japanese), *Journal of the Japanese Society for Quality Control, The 57th technical conference,* 53-60.

[14] AMASAKA, K., KIDO, T., MORIOKA, Y., and KAASAI, M.,(1997), A factor analysis of "CHIRASHI" adversity effectiveness, (in Japanese), *Journal of the Japanese Society for Quality Control, The 27th annual technical conference,* 35-38.

Chapter 5: "Science SQC", New Quality Control Principle

A New Principle Science SQC:

Proposal and Implementation of the "Science SQC", Quality Control Principle

In forecasting operation of the manufacturing industry in the 21st century, the authors recently proposed "Science SQC" as a demonstrative-scientific methodology and discussed its effectiveness on the basis of verification studies conducted by Toyota Motor Corporation. This study outlines a new SQC principle "Science SQC", as a demonstrative-scientific methodology, which enables the principle of TQM to be improved systematically.

Keywords: "Science SQC", a demonstrative-scientific methodology, to improve the principle of TQM, the principles of Toyota's quality control, the "next generation TQM (TQM-S)".

1. Introduction

To promote quality control that contributes to the world in the future, it is necessary for us to carry on lucid and reasonable TQM activities that will enhance the business process of all departments. To do this, it is important to give thought to quality control of the manufacturing industry in the future, change the principle of TQM activities accordingly and show a good example so that a brighter future may be obtained.

In this connection, the authors have proposed "Science SQC" as a demonstrative-scientific methodology and discussed the effectiveness of this method which improves the systematic development of the principle of TQM (Amasaka [1]to[3]). This paper positions the proposed "Science SQC" as the "next generation TQM (TQM-S)" that improves the principle of TQM and verifies its effectiveness through development at Toyota and subsequent results.

2. Needs for New SQC to Improve the Principle of TQM by Manufacturing Industries

2.1. Delay in Systemization of Quality Management System That Improves Manufacturer's Management Technology

It is generally agreed that quality management activities have contributed largely to Japan's economic prosperity today. Quality management by manufacturers originated in Japan when Statistical Quality Control (SQC) was introduced, used and deployed by Walter A. Shewhart [4] and W. Edwards Deming [5], who proposed that "quality management began and ended by control chart" [6], which is the basis of "quality built into process" [7]. These activities and results were advanced by J.M. Juran [8][9], systematically advancing the concept and progress method of the company-wide Total Quality Control (TQC) activities. This TQC has further advanced to today's Total Quality Management (TQM) activities [10].

In the 1980s and after, U.S. companies were stimulated by introductions made by Andrea Gabor [11] and Brian L. Joiner [12] and a MIT [13] report to change the concept of quality from product quality to quality of customer's sense of value by learning the quality management system in Japan. There, they reviewed Japanese- style cooperative activities and the effect of SQC as a scientific approach and deployed the new quality management nationwide, while also receiving instructions from W. Edwards Deming.

In the 1980s and 90s, however, Japanese manufacturers were too caught up in the economic boom both in Japan and overseas countries that they did not necessarily establish management technology bases sufficiently to prepare for the next generation. They put too much emphasis on JIDOKA (automation) introducing large-scale equipment and taking a long period of time and large investment for completion, transferring the production system accordingly. As a result, the production system became non-profitable. Some companies put too much importance on automatic adjustment with equipment having automatic adjustment functions even if 5M-E (Man, Machine, Material, Measuring and Method-Environment) deviate. In others, control charts simply disappeared from their manufacturing processes, reducing the scientific process management level.

As the situation continued, it became increasingly difficult to check the workmanship of products currently under production in real time. Not only for the production division, but also for the product design, production engineering and quality assurance divisions, workmanship became difficult to confirm. It seemed that the process maintenance, management and improvement cycle did not turn well, as the skill and problem solving capability of workshops, supervisors and auditors that could be improved through "observation of actual products" and "processes built into processes" had been abandoned for the reasons described above [14].

As alerted by T.Goto [15], Japanese manufacturers forgot the origin of quality management in the latter half of 1980s toward the early 1990s, resulting in a slowdown in growth. The most important factor was the delay in systemizing next-generation type management technology to be developed and introduced in accordance with technical advancements and changes occurring in the management environment around each manufacturer, making the whole Japanese industry fail to establish a new method (reasonable and scientific method for quality management) capable of improving the management technology. These are discussed in more detail by K.Yoshida [16] and K.Amasaka [17].

2.2. Necessity of New SQC as Demonstrative Scientific Methodology

The secret of successful quality control activities on the part of the manufacturing industries aimed at providing customers with attractive products consists of a reasonable way of thinking about quality control and the actual procedure to be established and followed accordingly. To be more precise, it means correctly converting customers' wishes (tacit knowledge) into engineering terms (explicit knowledge) by using the correlation technique, etc., replacing it with well-prepared drawings, and enhance the process capability for early embodiment into products.

In retrospect, the transition of quality control that developed from the manufacturing industry initially started with application of the mathematical method of SQC. It then developed into TQC using control technology, and more recently, into TQM using various management control techniques comprehensively. And the concept of quality has been undergoing expansion from conventional product quality-orientedness to business process quality, before becoming management technique quality-oriented. Along with this, the area of quality control activities expanded.

For the generator, the mentor and the promoter, it is becoming more difficult to bundle individual or workshop techniques of the total hierarchies or departments through existing quality control activities based on past successes that depended on proprietary rule of thumb or empirical techniques. Therefore, to produce really attractive products that satisfy customers, that is, to produce customer-oriented products, common terms (Management Technology: Quality Management System) become necessary for the circular business process cycle of all departments to correctly turn from sales, service, planning, development and design, to production engineering, manufacturing and logistics. As far as the writer knows, however, no SQC application system nor demonstrative methodology for effective renovation of the business process as a langue common to all departments has been seen for the new TQM activity for the coming age.

In this connection, Amasaka & Osaki [18], at Toyota proposed "Science SQC" as a demonstrative scientific methodology for the three core techniques (TMS, TDS and TPS) that form the principle of TQM activities shown in Fig. 1 to be linked organizationally. They also propagate and develop systematic and organized quality control. This represents the systematization of a new SQC application for creating technology by finding solutions using "general solutions" that can be generalized as "the principles of Toyota's quality control". They position this as the "next generation TQM management (TQM-S)" that enhances the centripetal force of Toyota's TQM activities. They further discuss the effectiveness of this SQC (Amasaka, et al. [19]).

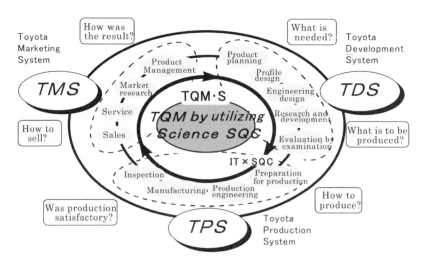

Fig.1. TQM Activities in ToyotaFig.1. TQM Activities in Toyota

3. Proposal and Implementation of the "Science SQC" Quality Control Principle

For today's demonstrative research, it is necessary to implement customer science that converts customer's wishes into engineering terms (Amasaka, et al. [20-21]). It is important for all departments concerned to share the objective awareness and turn tacit knowledge on business processes into explicit knowledge through coordinated activities.

The demonstrative-scientific methodology for realizing this conversion is called "Science SQC" (Amasaka [3]), in which SQC is utilized systematically and organically under a new concept and procedures so as to allow the four cores shown in Fig. 2 to mutually build on one another. This conceptual diagram shows the "Quality Control Principle" which forms the nucleus of TQM activities at Toyota as shown in Fig. 1 (Amasaka [3]). For details on the demonstrative research, refer to references (Amasaka [22]).

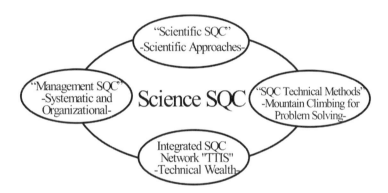

Fig.2. Schematic Drawing of "Science SQC"

3.1. Practical Application of "Scientific SQC"

The primary objective of SQC as applied by the manufacturing industry is to enable all engineers and managers (hereinafter referred to as businessmen) to carry on excellent QCD (Quality, Cost and Delivery) research activities through insight obtained by applying SQC to scientific and inductive approaches in addition to the conventional deductive method of tackling engineering problems.

It is important to depart from mere SQC application for statistical analyses or trial and error type analysis, and to scientifically use SQC in each stage from problem structuring to goal attainment by grasping the desirable form (Amasaka [1]).

3.2. Establishment of "SQC Technical Methods"

For solving today's engineering problems, it is possible to improve experimental and analysis designs by using N7 (the New Seven Tools for TQC) and the basic SQC method based on investigation of accumulated technologies. Moreover, it is possible to mountain-climbing problem solving by using a proactive combination of design of experiments as may be required using multivariate analysis amalgamated with engineering technology. The methodology in which the SQC method is used in combination at each stage of problem solving has spread and established as "SQC Technical Methods" (Amasaka [1]) for efficiently improving the jobs of businessmen.

3.3. Construction of Integrated SQC Network "TTIS"

Various cases of successful SQC application to actual business need to be systematized in order for them to contribute to forming engineering assets and help inheritance and further development. This is a prerequisite for the development of "Science SQC". This methodology is achieved with the "Toyota SQC Technical Intelligence System ("TTIS"), an integrated SQC network system that supports engineering problem solving. The "TTIS" is an intelligent SQC application system consisting of four main systems integrated for growth by supplementing one another (Amasaka [2]). For further details, refer to the references (Amasaka [23]).

3.4. Recommendation of "Management SQC"

The main objective of SQC in the manufacturing industry is to support quick solution of deep-rooted engineering problems. Therefore, the main objectives of "Science SQC" is to find a scientific solution for the gap generated between the theories (principles and fundamental rules) and reality (events). Especially in the application of "Science SQC", the differences from the principles and rules in an engineering problem should be scientifically analyzed to clarify six gaps that occur between theory, calculation, experiment and actual result to obtain a general solution. Filling these gaps results in an organizational problem.

For problem solving, it is necessary for the planning, design, manufacturing and marketing departments to clarify the six gaps, in other words to turn tacit knowledge on the business process to explicit knowledge for good understanding and coordination among the departments (Amasaka [3]). The methodology for organizationally managing the development of "Science SQC" is called "Management SQC". Recently, Amasaka et al. [24] further discussed and studied "Management SQC" through demonstrative cases of "Total Task Management Team activities" at Toyota Motor Corp. Fig. 3 shows a conceptual drawing of "Management SQC" for "Science SQC" development.

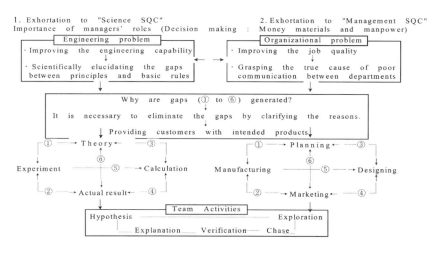

Fig.3. "Management SQC" for "Science SQC" Development

4. Demonstration Cases of "Science SQC" at Toyota

In view of recent changes in the environment surrounding manufacturing industries, enterprises are once again required to make efforts in further enhance the technology development capabilities of engineers. To do this, the need for SQC as a behavioral science must be newly recognized. SQC must be applied from a scientific standpoint for not only solving apparent engineering problems but also for foreseeing latent engineering problems, thereby contributing to the development of new technologies.   The following subsections describe the demonstrative cases of "Science SQC" as it is applied at Toyota.

4.1. Consistency in "SQC Promotion Cycle"

  Development of the "SQC Promotion Cycle" (Implementation, Practical Results, Education, and Growth of Human Resources) illustrated in Fig. 4 is essential for continuously increasing the expectations and effects of SQC applications. Amasaka [2] utilize this cycle as the principle for promoting "Science SQC" at Toyota. Practically, the SQC promotion cycle means a spiral activity where SQC is used for challenging today's engineering problems to enhance proprietary and management technologies.

  This will contribute to improving technologies, and subsequent results are reflected on hierarchical and practical SQC education to expand human resources, who in turn reflect the new technology in the performance of their operations. This mode of promotion constitutes the essence of "Science SQC". The figure schematically illustrates the concept of the "SQC Promotion Cycle" at a manufacturer (Amasaka [3]).

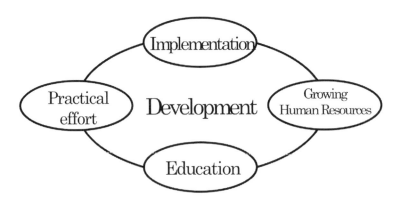

Fig. 4. Schematic Drawing of "SQC Promotion Cycle"

4.2. Systematization of Hierarchical SQC Seminar and Growth of Human Resources

To make the "SQC Promotion Cycle" spiral as illustrated in Fig. 4, it is important to establish a system that allows companywide development of systematic and organizational new SQC education and the growing of human resources. Conventionally, SQC seminars used to provide education on SQC basics, reliability analysis and design of experiments for many years. But each of them was based on manual calculation and the curriculums were set up for individual SQC methods, dependent on analyses. As a result, a dilemma was experienced by some participants in that they were unable to put what they learned into practice.

It is important to improve the contents of seminars to make them practically applicable to SQC education. To keep abreast of the times, it is important to apply SQC personal computer software "TPOS" (Amasaka & Maki [23]; Total SQC Promotional Original Software) and give importance to experimental and analytical designs so as to prevent us from ending up merely conducting trial and error analyses. From the viewpoints of Sections 2 to 3, hierarchical SQC education as shown in Fig. 5 is systematized as follows, and a step-up SQC seminar is planned and implemented for six hierarchies of Beginner SQC,

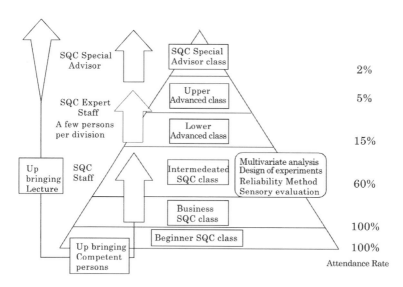

Fig. 5. Division of SQC Seminarian and Human Resources Training

Business SQC, Intermediate SQC, Lower Advanced, Upper Advanced, and SQC Advisor courses. The seminar attendant ratio is set up as illustrated in the figure for a total of 17,000 businessmen from all departments to ensure full growth of human resources (Amasaka & Osaki [18]).

The Beginner and Business courses are for all staff and the program is designed to provide enough training for them to carry on their routine business. The Intermediate course is for professionals who can freely apply the advanced SQC methods. Fig. 6 shows the curriculum established by Amasaka et al [19], for the Advanced SQC course (for engineering staff). During the initial period, they learn the concept and procedures of "Science SQC" at Toyota (SQC Review 1 to 5; SR1-5) from the instructors for the SQC advisory staff (advanced course). Each participant tackles individually registered themes (problem structuring and setting) and reflects what is learnt in their business.

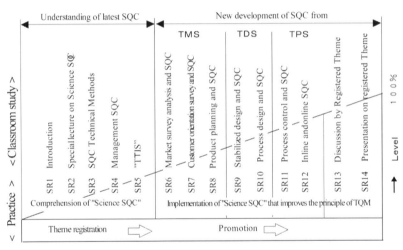

Fig. 6. Curriculum for Advanced SQC Course (12-day course)

The second period is for experiments and analysis designs which are important for maximizing problem solving. Lectures are given by an SQC level advisor of managerial position. The participants learn TMS (SR6-8) required for implementing "Science SQC" to improve the principle of TQM, and study demonstrative cases of "Science SQC" that help construct TDS (SR9-10), and TPS (SR11-12). In SR13 and SR14 in the third period, group discussion of registered themes takes place with the participation of instructors. This offers the opportunity for mutual study where the participants present their registered themes and verify the problem solving process.

Students having finished the advanced course and higher are qualified to be SQC special staff or SQC specialist advisors (approx. 1100 persons up to 2000). Under this system, they can display their ability as promotion leaders for company-wide "SQC Promotion Cycle" development. Recently, Amasaka[19, 25] set up a new 5-year course (1996 onward; for a total of 3000 engineering managers from all departments) for all managerial personnel (from directors down to divisional, departmental and sectional managers; approx. 2200 persons up to 2000) in an effort to strengthen the development of "Science SQC" for improving the principle of TQM.

4.3. Quality Improvement of Drive System Unit - Sealing Performance of Oil Seal for Transmission -

This is a typical example where engineers, who completed the new SQC training courses explained in Sections 4.2, challenged under instructions by the SQC special staff and advisors (Amasaka, et al. [26, 27]).

4.3.1. Engineering Problem

The oil seal for the drive system works to seal lubricant inside the unit. The cause and effect relationship between the oil seal design parameters and sealing performance is not necessarily fully clarified. As a result, oil leakage from the oil seal is not completely eliminated, presenting a continual engineering problem.

4.3.2. Conventional Approaches to Quality Improvement

So far, oil seal quality improvement has been made as follows. A development designer having empirical engineering capability recovers the leaking oil seal parts from the market, analyzes the cause of leakage with proper technology and incorporates countermeasures into the design. Many of the recent leaking parts, however, exhibit no apparent problem and the cause of the leakage is often undetectable. This makes it difficult to map out permanent measures to eliminate the leakage.

4.3.3. "Total Task Management Team" Activities

It is necessary to clarify the engineering problems to be tackled by sharing the essence of the problems as a team by combining the empirical technology of each individuals. Therefore, to explicate the essential engineering problems, it is necessary to improve the engineering problems generated by the lack of information among the related organizations (tacit knowledge), and their business processes. To do this, "Total Task Management Team"   (Amasaka, et al.[24]) methodology which was successfully used to improve brake performance quality is employed.

To be more precise, a total task management team named "DOS-Q5" (Drive-train Oil Seal Quality 5 Team) was organized in which the oil seal manufacturer participated. Fig. 7 is a Relational Chart Method showing the outline of the quality assurance (QA) team's activities. The QA teams in the figure are QA1 and 2 in charge of inquiries into the cause of oil leaks and design engineering problems, and QA3, 4 and 5 which handle manufacturing problems relating to the drive shaft, vehicle, and transaxle. QA1 and QA2 represent collaboration type team activities joined by the oil seal manufacturer (responsible for research, design, manufacture and quality assurance) which carries on quality improvement of the oil seal part.

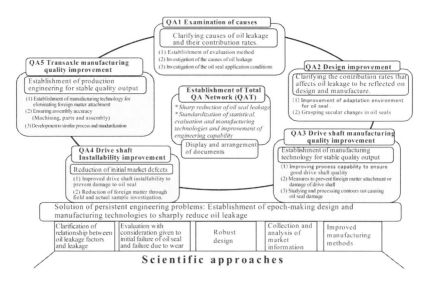

Fig.7. Relation Chart Method of Activities by "Total Task Management Team"

4.3.4. Implementation of "Management SQC"

To utilize proprietary intelligent technology and the insight of the team members (consisting of generators, mentors and promoters) by applying "Science SQC" as shown in Fig. 2, the core technology of "Management SQC" is implemented. To do this, it is found that the total quality assurance activities require further upgrading. Accordingly, Toyota's "Total QA Network" is being developed based on actual records of solving of this type of problem (Amasaka, et al. [24]).

Moreover, to promote overall optimization by maximizing the efforts of individual teams QA1 through QA5, problem solving is formulated using the "SQC Technical Methods" shown in Fig. 8. To realize optimization of the "DOS-Q5" business process, the "Management SQC" concept is applied in implementing information, production and engineering management.

(1) Improvement of Failure Analysis Process Using "Information Management [IM]"

The recovery method for oil leaking parts, the investigation method for oil leakage information and other failure analysis processes are devised to explore the true cause of oil leaks. For example, "non-leaking parts" are recovered together with leaking parts and subjected to discriminate analysis as shown in Fig. 9 and other cause and effect analysis. As a result, new knowledge is obtained indicating that the hardness of oil seal rubber affects oil leakage. Moreover, Weibull analysis using DAS (Sasaki [28], Toyota Dynamic Assurance System for failure analysis) reveals a new fact, that is, a mixed model consisting of three types of oil leakage (initial, random and wear) according to the running distance is acquired.

(2) Visualization of Oil Leakage Mechanism Using "Technology Management[TM]"

To study the oil leakage mechanism, a device is developed to visualize the dynamic behavior of the oil leakage from the oil seal lip section. Through factorial analysis of the collected data using multivariate and experimental analyses employing a new experimental design, the oil leaking mechanism is visualized and the following causal analyses are conducted:

(a) Association of running distance, lip surface roughness and pumping volume [new fact].

(b) Association of the inside diameter of the differential case, the lip wear width and the roughness of the axis [new fact and quantification].

(c) Association of roughness of the axis and pumping volume [new fact].

(d) Association of differential case wear and driveshaft eccentricity [new fact].

(e) Association of the lip tightening margin and lip wear [quantification].

As the result, the above-mentioned associations are explicated. New knowledge is acquired including the quantification of conventional empirical technology and the creation of new technology not found in the empirical technology. The oil leaking mechanism shown in Fig. 10 is thus estimated.

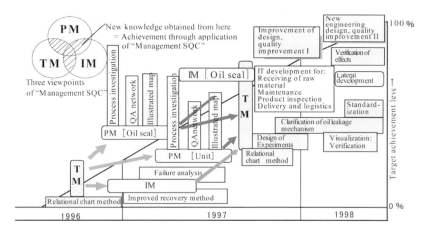

Fig. 8. Problem Solving by Utilizing the "SQC Technical Methods"

(3) Development of "Total QA Network" Using "Production Management [PM]"

For example, to improve the quality of the oil seal parts by incorporating the above-mentioned IM and TM knowledge, QA1 and 2 analyzed and stratified the types of oil leakage and problems in the manufacturing processes, then developed illustration mapping by making relative analysis of these two elements. The analytical results are reflected on the "QA Network Table" (Amasaka, et al. [20]) used to develop process control into a science (visualize).

The business process could be visualized from receiving to delivery inspection and logistics (distribution). Next, QA3, 4 and 5 teams similarly developed the "QA Network" for the drive system manufacturing process. Then they established the process control science all the way from receiving to delivery. With the coordination of QA1 and 2, they could clarify the fact (visualization of behavior) that, for example, oil leakage occurs if foreign matter the thickness of a hair (75 (m)) is attached to the drive shaft.

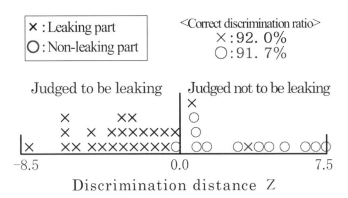

Fig. 9. Result of discriminate analysis

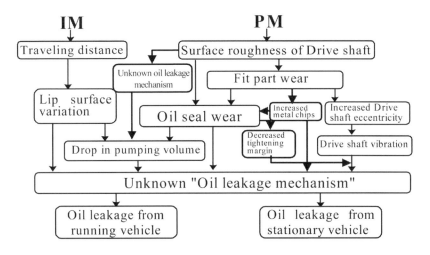

Fig. 10. Estimated Oil Leakage Mechanism

(4) Achievement of "DOS-Q5"

Through implementation of "Science SQC", the oil seal leaking mechanism whose cause was unknown has been scientifically visualized and verification obtained using cause and effect analysis, with more proprietary knowledge of the technology subsequently acquired. We thus eliminated the oil leakage problem and achieved our target.

5. Conclusion

Demonstrative and scientific methodology is established for SQC applications aimed at improving technology. "Science SQC" is proposed as a new, systematic and organizational SQC application methodology for the manufacturing industry. Toyota Motor Corporation has demonstrated that this methodology can improve the quality of work of engineers in every stage of their business process and contribute to creating products of excellent quality. In the future, "Science SQC" will be positioned as a quality control principle and applied to solving various practical problems. And with the accumulation of demonstrative studies, "Next Generation TQM (TQM-S)" designed to improve the principle of TQM will hopefully be established.

References

[1] K. Amasaka,(1998), Application of Classification and Related Method to the SQC Renaissance in Toyota Motor, *Data Science, Classification and Related Methods*, 684-695, *Springer.*

[2] K. Amasaka,(2000), A Demonstrative Study of A New SQC Concept and Procedure in the Manufacturing Industry -Establishment of A New Technical Method for Conducting Scientific SQC-, *An International Journal of Mathematical & Computer Modeling*, 31(10-12), 1-10.

[3] K. Amasaka,(1997), The Development of New SQC for improving the principle of TQM in Toyota -from "SQC Renaissance" to "Science SQC"-, *Proceedings of the CSQC Conference and the Asia Quality Symposium, Tainan, Taiwan*, 429-434.

[4] Walter A. Shewhart,(1986), Statistical Method from the Viewpoint of Quality Control, *Edited and with a New Foreword by W. Edwards Deming, Dover Publications, Inc., New York.*

[5] Walton. Mary,(1988), The Deming Management Method, *Dodd, Mead & Company, Inc., New York.*

[6] William B.Rice,(1947), Control Chart in Factory Management, *John Willy Sons, Inc., New York.*

[7] I. Kusaba (Supervisor): M. Kamio, K. Amasaka et al., (1995), Fundamentals and Applications of Control Charts,(in Japanese), *Japanese Standards Association.*

[8] J.M.Juran,(1989), Juran on Leadership for Quality -An Executive Handbook, *The Free Press, A Division of Macmillan, Inc.*

[9] J.M.Juran,(1988), Juran's Quality Control HANDBOOK, *McGraw-Hill, Inc.*

[10] Edited by TQM Commitee,(1998), Overall Quality Management in TQM 21st Century, (in Japanese), *Union of Japanese and Engineers.*

[11] Andrea Gabor,(1990), The Man Who Discovered Quality :*How W.Edwwards Deming Brought the Quality Revolution to America, Originally published by Random House.*

[12] Blian L.Joiner,(1994), Forth Generation Management: *The New Business Consciousness by Joiner Associates, Inc.*

[13] Dartouzos, M.L,(1989), Made in America, *MIT Press.*

[14] K.Amasaka,et al.,(1999), Apply of Control Chart for Manufacturing - A Proposal of"TPS-QAS" by In line-On line SQC-, (in Japanese), *Journal of the Japanese Society for Quality Control, The 29th Annual Technical Conference,* 109-112.

[15] T.Goto,(1999),Forgotten Origin of Management-Management Quality Taught by GHQ, (in Japanese), *CSS Management Lecture, Productivity Pub.*

[16] K.Yoshida,(1999), International Strategy Learned from Restoring U.S.A., (in Japanese), *Union of Japanese Scientists and Engineers, The 68th QC Symposium,* 61-65.

[17] K.Amasaka,(1999), The TQM Responsibilities for Industrial Management in Japan-The Research of Actual TQM Activities for Business Management-, (in Japanese), *The Japan Society for Production Management, The 10th Annual Technical Conference,* 48-54.

[18] K. Amasaka and S. Ozaki,(1999), The Promotion of New SQC Internal Education in Toyota Motor -A Proposal of "Science SQC" for Improving the principle of TQM-, *The European Journal of Engineering Education,* 24(3), 259-276.

[19] K. Amasaka, et al.,(1998), A Study of the Future Quality Control for Bring-up of Businessmen at the Manufacturing Industries, -The significance of SQC study abroad for students in Toyota Motor-, (in Japanese), *Journal of the Japanese Society for Quality Control, The 60th Technical Conference,* 25-28.

[20] K. Amasaka, et al.,(1999), Studies on "Design SQC" with the Application of "Science SQC"-Improvement of Business Process Method for Automotive Profile Design-, *Japanese Journal of Sensory Evaluation,* 3(1), 21-29.

[21] K. Amasaka, et al.,(1998), The Development of "Marketing SQC" for Dealers' Sales Operating System, -For the Bond between Customers and Dealers-,

(in Japanese), *Journal of the Japanese Society for Quality Control, The 58th Technical Conference*, 155-158.

[22] K. Amasaka,(1995), A Construction of SQC Intelligence System for Quick Registration and Retrieval Library, -A Visualized SQC Report for Technical Wealth-, *Lecture Notes in Economics and Mathematical Systems*, 445, 318-336, *Springer.*

[23] K. Amasaka, and K. Maki,(1992), Application of SQC Analysis Software at Toyota, (in Japanese), *QUALITY, Journal of the Japanese for Quality Control, The 25th Annual Technical Conference*, 3-6.

[24] K. Amasaka, et al.,(1997), The Development of "Total QA Network" by utilizing "Management SQC",-Example of Quality Assurance Activity for Brake Pad-, (in Japanese), *Journal of the Japanese Society for Quality Control , The 55th Technical Conference*, 17-20.

[25] K.Amasaka.et al,.(1999). A Proposal of the New SQC Education for Quality Management - A Propose of "Science SQC" for Improving the Principle of TQM -,(in Japanese), *Quality, Journal of the Japanese Society for Quality Control*, 29 (3), 6-13.

[26] K. Amasaka, et al.,(1998), A proposal "TDS-D" by Utilizing "Science SQC" -An Improving Design Quality for Drive-Train Components-, (in Japanese), *Journal of the Japanese Society for Quality Control, The 60th Technical Conference*, 29-32.

[27] K. Amasaka, and S. Osaki,(2002), A Reliability of Oil Seal for Transaxle - A Science SQC Approach in Toyota -, *Case Studies in Reliability and Maintenance by Wallace R. Blischke and D. N. P. Murthy, to be published by John Wiley & Sons, Inc.,* ISBN 0-471-41378-3, 571-581.

[28] S. Sasaki,(1972), Collection and Analysis of Reliability Information in Automotive Industries, (in Japanese), *Union of Japanese Scientists and Engineers, The 2nd Reliability and Maintainability Symposium*, 385-405.

3. Human Resource Development and Practical Outcomes of Science SQC in Toyota

Chapter 6: "SQC Promotion Cycle" Activities

Improving Human Resource Development:

SQC Development and Effects at Toyota

If we take a look back upon the history of QC development, the current flourishing of TQM is no doubt attributable to the important role played by the foundation of SQC. Now, however, in the midst of recent waves of notable technological innovations, a new role for SQC is beginning to be called for throughout the entire manufacturing process, from downstream manufacturing to planning, development, and design at the source.

The issues facing companies today are acquiring the latest SQC techniques and taking up the challenge of new issues. This paper, "SQC Development and Effects at Toyota," will present some of the ways in which SQC is used to deal with current issues at Toyota today in hopes for a rebirth of SQC for the new era.

Keywords; SQC Training, Human Resource Development, Practical Outcome, SQC Renaissance, Toyota.

1. The SQC Renaissance and Its Deployment

In recent years, Toyota has been deploying a "SQC Renaissance[1]" company-wide in order to improve the quality of work of technical staff and efficiently promote problem solving.

The objectives of this program are to improve the quality analysis capability of all staff members, to effectively utilize SQC in practical issues such as new technology, new manufacturing methods, and bottleneck technology, and to produce practical outcomes that have a high degree of universality and continuity.

The promotional measures to realize these objectives are (1) create and establish a promotional organization, (2) train personnel to enrich and support that organization, and (3) enhance and maintain SQC training, which will become powerful nourishment to promote the organization and personnel.

The authors, fortunately, are in a general position of planning and administration for this deployment activity. Following is an outline of the distinctive matters that have been considered and implemented on a daily basis as part of the three promotional measures described above during the several years in which this activity has been deployed. This paper focuses mainly on front-line technical staff with the rank of assistant manager.

2. Freshly consider and implement SQC

2. 1. The two "pillars" of business management technology: TQM and TPS

In order to manufacture well vehicles that customers will purchase enthusiastically, timely QCD research is important. To achieve this, the two business management technology "pillars" of TQM and TPS (Toyota Production System) are the foundation of manufacturing at Toyota[2].

As shown in the conceptual diagram in Fig. 1, these two pillars interact with one another in a software-hardware relationship in terms of elements such as Q, C, and D. They consistently play an important role in flattening out extreme wave patterns into small fluctuations and raising average values.

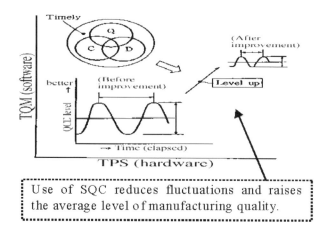

Fig.1. Relation between TQM and TPS

2. 2. SQC is the origin of TQM

TQM, in a word, is a company-wide activity that developed from SQC before TQM even existed. Equipped with the tool of SQC and while spiraling up, it is the practice of habitually asking "What improved and how much? Will it improve?" to maintain, perform kaizen, and raise consciousness. SQC constantly delivers nourishment to the "timber" of TQM, which is the backbone of manufacturing, and plays an important role in helping TQM bear fruit. In that sense, it may be simply said that SQC is the origin of TQM.

2. 3. The challenges of SQC today and SQC implementation

For the purposes of mutual study and human resource development of SQC, Toyota inaugurated a Toyota-wide SQC Seminar in 1967. Documents and cases were studied, and the results were published in many SQC manuals[3] and other materials which were issued to participating companies and were of great use in their training and actual work.

In the midst of the transformation of the era from learning from the West to manufacturing by itself, in recent years Toyota has begun to hold regular SQC Symposia in order to promote future QC throughout Toyota. The issue of how SQC should be implemented is studied intensively from various aspects. In taking excerpts from and summarizing these, the following issues are raised.

The first issue is the diversification of the issues we must address, such as grasping customer needs, quality and sensitivity, and the movement from ppm management to Fit. The mere simple application of conventional SQC techniques will not be fully recognized as a powerful tool for data analysis by engineers.

For the purposes of problem solving for engineers, what is called for is a change from a simple tool to a new technique development that will become a powerful weapon for overcoming new hurdles. In addition, at the same time, due to the growing proliferation of computers, the development and utilization of extensive software for SQC techniques that will make data analysis more efficient is necessary.

For the second issue, let us look at an excerpt from a symposium with the theme of "Let's utilize SQC more to improve work quality." Realistic issues were raised from the respective standpoints of the QC Promotion Secretariat, SQC specialists, workplace supervisors, and people in charge of actual work; these are summarized in Fig. 2. The issue common to all areas-SQC training, implementation, and outcomes-was a lack of human resources capable of instructing SQC, or instructors.

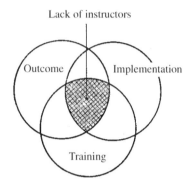

Issues

(1)Planning and administration that is useful for actual work
    (matching trends of times)
(2) Creation of scheme to improve practical outcomes (continuity)
(3) Creation of scheme to promote human resource development (expansivity)

Fig. 2. Toward promotion of SQC that is useful for actual work (from Toyota-
    wide SQC Symposium)

In the area of planning, a scheme (planning and administration) that matches the trends of the times is important, and the creation of a system to raise practical outcomes and develop human resources is the responsibility of the instructors. In addition, in the area of implementation, the training of technical staff proficient in SQC that can give hands-on instruction for concrete problems is important and will become nourishment for the future.

In response to the two issues summarized above, the SQC Renaissance that Toyota is promoting is a company-wide SQC rebirth program equipped with promotion circles as shown in Fig. 3. It is not simply a Toyota activity, but an activity to promote SQC that has repercussions throughout the Toyota Group[4].

The details of the activity will be discussed in "3. SQC Training: Planning and Administration". Of the matters that are actually being undertaken now in response to the many issues above, the area of SQC training is the main focus of the following pages. "Improving Human Resource Development and Practical Outcomes," which is the subtitle of this paper, is addressed as well.

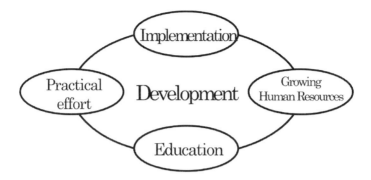

Fig. 3. Schematic Drawing of "SQC Promotion Cycle"

3. SQC Training: Planning and Administration

Within a company, the objective of SQC training should not be merely to learn SQC. Its true value lies in how much it can be reflected in actual work outcomes. The first point to be noted in planning and administration of SQC training is the construction of SQC seminars that are useful to various types of work in each division from the source to downstream. That includes the writing of a hands-on curriculum that enables SQC tools to be effective in practical application and the enrichment and maintenance of various courses that lead from the beginner to the advanced level with clear objectives.

The second point of SQC training is the enrichment of training quality that matches the era and provision of elaborate training materials and tools that are useful for efficient work procedures. The third point is the human resources development of a technical staff that reaches several thousand in number. This involves the creation of a training system that ensures improvements in the skills of seminar participants (quality) and maintains the number of the participants (quantity) continuously and systematically. The fourth point is a follow-up system that will (i) encourage practical improvement by those who have completed the seminar to improve and (ii) enable checking of practical outcomes and achievement in order to utilize it for future improvement of training. The fifth point is the qualitatively and quantitatively enhanced establishment of instruction staff. It is especially important to promote the training of instructors continuously according to schedule. As mentioned below in "3. 2 Human resource development" and "3. 3 Actual implementation," it is important to create a system that will enable the sufficient acquisition of instructors who are strong in both SQC and practical work.

3. 1. Maintenance and enhancement of stratified SQC training

In the past, Toyota offered the SQC introductory course, specific technique courses for reliability and design of experiment, and the Toyota-wide SQC Seminar to which trainees who completed those courses were selected and dispatched. In recent years, with the objective of fundamentally revising the SQC training system, progress has been made in the maintenance and enhancement of stratified (5-level system: foundation-beginning- intermediate-advanced-most advanced) SQC training as shown in Fig. 4[1][5].

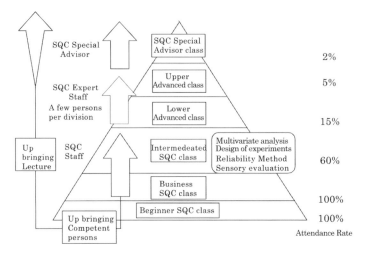

Fig. 4. Division of SQC Seminarian and Human Resources Training

(1) New establishment of SQC foundation course (for new employees)

In addition to TQM training in the orientation period, training for introductory level SQC techniques was established. All new employees receive four-day introductory training in one of several sessions held in the six-month period after being posted to a new workplace. This aims to further the widespread use of SQC among low-ranking employees, and the progress of the actual utilization of

SQC techniques can be heard whether in practical themes of practical training reports or workplace presentation meetings.

The instructors are all young technical staff from every division (employees who have completed the seminar from ranks 1-3); they take turns lecturing on actual cases from their own experience. This approach leads to both self and mutual study, and steady outcomes are visible in both human resources development and increased utilization of SQC in the workplace.

(2) SQC Beginning Course (SQC basic courses divided by technical field)

The former SQC introductory course was qualitatively enhanced and strengthened. It was transformed to transcend a handbook-style understanding of SQC to become a hands-on SQC basic course. It was expanded so that the better part of technical staff with at least two years of experience with the company can participate. In addition, it was transformed from classroom training that covered divisions from the source to downstream into specialized courses divided by technical field.

Reinforcement of instructors from both inside and outside the company and original textbooks were used for various purposes. Courses were dramatically enhanced and maintained in terms of both quality and quantity by means such as strengthening elements including automotive reliability and sensory evaluation to match the trends of the times. Looking at division by technical field, in the research and development divisions, courses that have a direct bearing on actual work, such as "How to determine the sample size required for an experiment" and "Nonparametric methods," were incorporated.

In the design and production engineering divisions, courses such as "Sampling" and "How to determine specifications well" are offered; in the production preparation divisions and plants, courses such as "How to do production preparation and process control well" are offered. One-fourth of the curriculum is hands-on in order to be useful to the applicable division. In addition, initiatives that lead directly to actual work outcomes have been enhanced and strengthened.

The first initiative was enabling people who register practical issues as themes to take courses. The second is the completion of a theme one year after registration and a system for workplace presentation meetings. In order to raise the theme completion rate and the achievement level, it is important to have trainees learn from experience how to practically and effectively utilize SQC techniques in each course. To achieve this, as the third initiative, with the guidance of the SQC specialist staff and SQC specialist advisors described later, using a group roundtable system, one-fourth of the curriculum consists of guidance meetings for registered themes or case seminars; these contribute to the improvement of training outcomes. The fourth is the addition of practical computer training courses that utilize the SQC original software described below. The utilization of this software is having an impact on work efficiency.

(3) SQC intermediate course (specialized courses in design of experiment, reliability, multivariate analysis)

  Both of the specialized courses, design of experiment and reliability, were differentiated qualitatively from the beginning course and arranged as a curriculum for the intermediate course. The intermediate course is one level above the beginning course; it also includes the new course, multivariate analysis. These courses are available to those who completed the beginning level course. As more than half of the beginning course trainees progress to the intermediate level, the numbers of trainees doubled in the three courses.

1) The Design of Experiment Specialized Course was enriched with split-plot design, direct sum method, direct product method, and parameter design, which are effective for experiment efficiency and in the area of practical outcomes. Furthermore, based on the comments of past participants, the course was changed from one that over-emphasized analysis to one that focuses on methods of designing experiments and technical analysis of experiment results. Furthermore, like the beginning course, there are a number of experiments to heighten actual results.

2) Under the concept of the future reliability of Toyota, the Reliability Specialized Course aims to enhance reliability analysis in order to pinpoint the area of the hem of the distribution from the orientation of the average value of a lifespan distribution. In addition, in order to cover the area of theory and practice in reliability and enter the area in between deeply, we have invited the participation of many experts from divisions of origin to downstream process divisions as instructors. Like the beginning course, this course was changed to a course for intensive study of the way Toyota should be along with the trainees through group roundtable discussions on practical research cases or registered theme guidance meetings.

3) Furthermore, the distinctive thing is that a new Multivariate Analysis Specialized Course has been established. More than 200 people have already finished the course in the past few years, and a great deal of case research and outcomes that have a direct bearing on actual work have become evident. Parts of these are presented in the form of extra-company presentations at the QC conferences and other gatherings, contributing to the human resource development of young employees. This course is conducted with one computer per trainee, and there are a small number of trainees per course; the practical curriculum consists of lectures and practice. The materials comprise original software developed by Toyota and textbooks that make use of a large number of actual cases from each division.

The main analysis techniques incorporated into the course are analysis methods such as multiple regression analysis, principle component analysis, discriminant analysis, and clustering, and in addition to quantification theory I-IV and factor analysis, time-series analysis, curvilinear regression analysis and others. Furthermore, in comprehensive supplementary classes, trainees learn how to view and interpret data and curve fitting techniques in the context of a specific technology.

The instructor staff, consisting of people of practical experience including university professors and the authors, are supplemented by trainees who have completed the seminar who act as instructors in rotation. The most distinctive feature of this course is that a multivariate consultation room is frequently held for the actual issues registered as a theme when all trainees signed up for the course.

The consultation room offers individual guidance to support and guide the progress of the registered themes from various angles. The impact of this can be seen to be extremely positive from looking at the proportion of themes completed and the achievement level. People who complete a registered theme and achieve practical outcomes participate in these group-discussion style seminars to present their achievement as case methods. They also participate in registered theme case presentation meetings, which are another component of this course, and this enables them to study intensively together with the other seminar participants.

(4) Advanced courses (SQC specialist staff workshop, SQC advanced course)

These courses were established in the last few years, aiming at human resources development that would constitute the core of each workplace and that would be necessary to the powerful deployment of "company-wide SQC promotion" (described later). The courses are aimed at people selected from those who have mastered the intermediate course, especially those recommended by divisional managers.

1) SQC specialist staff workshop

This course, in addition to utilizing SQC techniques from the advanced level, especially emphasizes practical ability. Not to mention inherent technology, it aims to train hardy instructors who can use SQC as a powerful tool. The class consists of a total of thirty to forty trainees from three areas of divisions: research and development/design/evaluation, production technology, and manufacturing. There are nearly ten instructors: the authors, the TQC Promotion Division, and managers and general managers from each division. It is a one-year course held once a month, including retreats twice a year.

The curriculum is divided into five parts. The first is learning advanced level SQC techniques and their application using examples from actual work. In the second, seminars with a system of individual research on excellent SQC research cases and group discussion meetings are held before and after. This course equips the trainee with guidance ability appropriate to a workplace instructor though self and mutual intensive study.

In the third part, trainees take up the challenge of the advanced level theme that was registered when they joined the course and try to solve it with a high degree of universality. Through group discussion and, as a matter of course, individual research, the qualitative completion level is high. In the fourth level, actual work issues are extracted in the homework and exercises, for example in multivariate analysis, to offer the trainee first-hand experience in recognizing that the ability to read a multivariate association graph with computer software depends on the capability of analyzing it from specific technology aspects.

As the fifth part, evaluation of a thesis comprehension and final research presentation meeting for registered themes to which supervisors of the trainees are also invited is incorporated into the course to confirm the instructors' abilities. In this course, a training diagnosis chart is utilized as a record of contact to facilitate follow-up of the progress of training. It is passed from trainee to supervisor to instructor to the TQC Promotion Division, and its utility is great.

2) SQC advanced course

This course is based on the curriculum for the SQC specialist staff workshop. Like the beginning and intermediate courses, it is divided into five segments and lasts for twelve days spread over three to four months. The course emphasizes intensive study of theoretical aspects, and furthermore the areas covered by theory and practice and the area in between.

In the same way, many trainees, who are selected from the three divisions from origin to downstream process divisions, proceed to the most advanced course described below. Ten-odd first-rate university researchers are invited to be instructors; in addition, engineers of the rank of manager or above who have a mastery of SQC participate as in-house instructors. Instruction in each section, lectures, and case research seminars based on the group discussion method progress jointly. In addition, there is a new system in which trainees spend one year after the completion of the course, doing graduate research on the theme that they registered when joining the course. They make presentations on the results at the next seminar.

The curriculum makes use of advanced level research reports and Toyota Group research cases. The categories covered in the curriculum include SQC specialist staff workshop, design of experiment, parameter design, multivariate analysis and time-series analysis, sensory evaluation methods, reliability techniques, utilization of SQC techniques when the number of N is small (nonparametric methods, etc.), and SQC in production/information.

Both courses have evolved from the more traditional SQC that tended to emphasize analysis and address the question of how to attack issues with an emphasis on the action on the planning stages. Throughout the courses, trainees gain actual experience in investigation methods as well as prediction and verification by identifying issues, setting objectives, creating solution guidelines, and selecting the action to take. The objective of this course is to boost the ability of workplace supervisors and develop the "brain" that constitutes the ability to guide.

(5) Most advanced course (Toyota-wide SQC training seminar system)

The Toyota-wide SQC training seminar described above was revised in recent years and positioned as the course that will provide human resource development for the Toyota Group of tomorrow. It consists of three working groups: design evaluation of research and development divisions, process design of production engineering divisions, and process control of manufacturing divisions. The class is composed of over sixty people, including both first- and second-year trainees: one from each division of each of the eleven Toyota Group companies.

Guidance advisors for each working group are people with practical experience of the rank of manager or above who have a mastery of SQC dispatched from the Toyota Group. The course meets once a month for two years, in two cycles of one year each. Each working group incorporates the lectures and guidance of a full-time university researcher. It is an enhanced seminar, with elements such as a presentation of registered themes by working group at the end of the academic year and joint presentation meetings for all three working groups, at which trainees can receive the guidance of the university researchers.

In order to advance to the next level or graduate, thesis intelligibility is evaluated with high-level SQC research papers to judge guidance ability and the presentation of the registered theme at the presentation meeting is assessed.   In this way, the system is designed to develop people of high caliber in both name and substance.

This course has three objectives. The first is for trainees to research actual, high-quality themes that have universality and continuity in order to become assets for their companies and divisions. The second is for the trainees to conduct intensive self and mutual study though revising and compiling the SQC Manual, which is the footprint left by the intensive study of their predecessors in this seminar, thereby creating an SQC asset for the Toyota Group. The third is for trainees to conduct research on high-quality themes that are useful for actual work as the powerful "brain" for promoting SQC in their respective companies and divisions after they graduate from the seminar. Furthermore, along with those who have completed the advanced course, they can bring out the guidance ability as SQC "spark plugs" in the workplace. We expect that they will broaden and enliven the future, whether inside or outside the company.

(6) Numerous experiments that are useful to actual work

1) Theme registration and workplace presentation meeting system of registered
   theme
   When trainees join the beginning course and above, they register an issue they are grappling with in their actual work as a theme. After the seminar, they complete the theme within a predetermined time period and present it in the workplace. The system focuses on reflecting this in practical outcomes.

2) System of study meetings to discuss attack method for registered theme

In order to accomplish this, in each course, the SQC specialist staff and SQC specialist advisor described below in the last paragraph sit down together with trainees in small groups to give guidance about how to attack registered themes in order to solve them. This system supports their supervisors in providing guidance on actual work when trainees work on their themes when they return to the workplace.

3) Case workshop system based on group roundtable system

In order to further the trainees' theme research and facilitate the practical utilization of SQC in actual work, a case workshop system based on group discussion was established in each course. The case workshop may take the form of theme presentations given by trainees of higher levels or it can be based on the actual examples and experience of SQC specialist staff and special advisors.

4) Utilization of computer to make actual work more efficient[3]

The original software TEDAS (Toyota Experimental Data Analysis System) and TFAS-M/T (Toyota Forecast & Analysis Support System Maltivariate / Time Series Analysis), using personal computers, have come to be used within Toyota and in other places. Linked with the recent proliferation and increased capabilities of personal computers, the Toyota TPOS (Total SQC Promotion Original Software) series of personal computer software is used to increase the efficiency of data analysis. Used widely in the practical training of stratified SQC seminars as well as in actual work, its utility is extremely high. Aiming at qualitative periodic enhancement and revision with the holding of each course, it is being provided sequentially to each workplace.

(7) Expansion of Toyota Group

In recent years, SQC training has been undergoing revision and enhancement at each Toyota Group company just as at Toyota. At ten group companies, especially, human resource developing is progressing; they are setting up beginning and other courses themselves, and quite a few people are participating in courses at the intermediate level and above in Toyota's stratified training. In addition, they are holding special courses jointly with other affiliated companies. Toyota is engaging in cooperation with and support for the eleven Toyota Group companies such as planning and administration and dispatching instructors. A new expansion is currently becoming evident.

3. 2. Stratified human resource development implementation

As described below in section 4. 3, it is necessary to continuously and systematically promote human resources enhanced both in quality and quantity utilizing SQC techniques to improve practical outcomes. In order to accomplish this, Toyota is establishing various human resource development systems with objectives such as the following.

The first, as shown in Fig. 4 in the previous section 4. 1, is to link stratified SQC seminars to practical utilization and encourage employees to participate in and master more advanced courses progressively and steadily.

The second objective is to have individuals and groups utilize SQC in their actual work to heighten their mastery of SQC. It is important that, along with improving their own skills, they guide their juniors to increase their guidance abilities. Learning the value of SQC from mistakes and successes and taking on the challenge of problem solving at ever-higher levels is nourishment for the development of human resources.

The third objective is to have trainees experience being instructors and advisors in stratified SQC training and SQC workshops and case seminars in the workplace. This will enable further self and mutual intensive study of the basics of SQC techniques and their application to actual work.

The fourth objective is for employees to attend research presentation meetings and symposia inside and outside the company to deepen their insight. Having the opportunity to give presentations and receiving instructions and advice from workplace advisors and SQC instructors is extremely beneficial. For young engineers, especially, giving presentations before people with a knowledge of SQC and getting their guidance is a good chance for them to revisit the basics of how to do their work, and provides significant nourishment for the future.

(1) Training for those who complete stratified SQC seminars

1) SQC staff system
"SQC staff" refers to those who have completed the beginning and intermediate courses. They conduct practical research on their own within their departments or sections. They take turns gaining experience in giving explanations of exercises

or homework in the foundation or beginning courses one or two levels below their own level, as well as act as advisors to group roundtables or give lectures. SQC staff includes many middle-ranking technical staff members, including assistant managers.

2) SQC specialist staff system

"SQC specialist staff" refers to those who have completed the advanced and most advanced courses. They conduct research on high-level practical issues at the divisional or larger organizational level, and provide guidance on practical themes to SQC staff.

Furthermore, they take turns gaining experience in acting as advisors to the foundation, beginning, and intermediate courses as well as give lectures. SQC specialist staff includes many upper-ranking technical staff such as assistant managers and section managers.

3) SQC specialist advisor system

"SQC specialist advisors" are outstanding managers at the rank of section manager or above. Their ability is at the level of the advanced or most advanced courses or even higher and they have conducted distinguished intensive study over many years of actual work. Furthermore, they have experience in acting as instructors of SQC seminars and giving research presentations inside and outside the company. They, of course, provide guidance to SQC staff and SQC specialist staff in the course of actual work, as well as display their guidance abilities in the planning and administration of promoting SQC that has a direct bearing on actual divisional work.

(2) Personnel registration and qualification system

In this personnel registration system, SQC specialist staff and SQC specialist advisors who receive recommendations from their workplace supervisors undergo a qualification examination as personnel who will become the core of each division's SQC promotion. Their qualifications are then certified at many strata. This system allows personnel to exhibit their leadership individually as well as at an organization level and will lead to systematic and continuous human resource development.

(3) Presentations at extra-company SQC research presentation meetings

As described below in section 3. 3 and shown in Fig. 5, in recent years, Toyota has been having employees, mainly young SQC staff, utilize SQC and actively make presentations outside the company on the research cases that bore successful outcomes in actual work[6]. In recent years, there has been a trend for extra-company presentations to increase. This signifies that the establishment and enhancement of stratified seminars and practical research in actual work are becoming established in the scheme of work. It is an example of SQC application that has a direct bearing on any practical outcome, and expansion of SQC can be seen from upstream to the origin level divisions.

This is the outcome of the advice and guidance of SQC specialist staff and SQC advisors, not to mention that of supervisors. Year after year, each workplace is moving in the direction of encouraging the active giving of presentations for purposes of human resource development. In the future, we hope that further opportunities for presentations will be given to SQC specialist staff to increase the presentation participation for QC and other engineering fields. We believe that such opportunities for human resource development will further encourage even more sophisticated practical research as nourishment for growth.

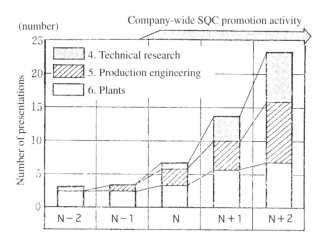

Fig. 5. Extra-company presentations of SQC research cases
(Target: QC conferences, QC-related bodies, other)

3. 3. SQC promotion project among company-wide staff[6-7]

The objective of the SQC Renaissance is the improvement in quality analysis capability of the technical staff. In the SQC Renaissance, promotion activity first started with structuring a promotion system at each division in production engineering and the plant divisions; technology divisions (research and development, design, evaluation) followed. The purpose of this activity was to promote high quality, timely QCD research activities with SQC with all technical staff members registering themes that have a bearing on actual work. The following is a brief outline of Phase 1 of the SQC promotion project that was carried out in each division over three years (1988-1990).

(1) Phase 1 activities

The first year of the promotion project was a time to prepare the soil and sow seeds in order to give birth to SQC workshops and practical outcomes. The second year was a time for germination and treading on the wheat plants for the purpose of establishing beginning and intermediate level SQC practical research and accessible problem solving. The third year was a time of growth for young leaves, a time for promotion of SQC utilization research that was effective in its application to actual work and the challenging themes to attain a higher level of the practical outcomes (described later).

(2) Devising a scheme for SQC promotion

The first element was the creation, maintenance, and enhancement of a promotion organization. The TQM Promotion Division became the general secretariat, and promotion organizations were established in each division category. Divisional secretariats were established, and plans for promotional measures were made in each division. In deployment to actual work, an SQC promotion committee chair (of the rank of divisional general or assistant general managers) and promotion committee members (SQC staff from each department) were appointed. A division-oriented planning and administration system was instituted.

The second element was the systematic participation in stratified SQC seminars, described in section 3. 1, for the training and fortification of the SQC staff who were to deploy the promotion. In the second and third years, the enhancement of promotion committee members was especially sought to build up the practical outcomes described in section 3. 2. A vice committee chair (section manager rank) was established to assist the SQC promotion committee chair, and the project was deployed to actual work with SQC specialist staff as the core. The third element was the promotion of theme registration and completion of registered themes. The target for the first year was completion of one theme per person per year. Divisions that had made progress in establishment, especially, encouraged their departments and sections to undertake technically important, challenging themes.

(3) Holding of challenging theme attack consultation room

  Furthermore, in the second and third years, it was decided to undertake critical issues, such as new technology, new production methods, and bottleneck technology, aggressively in each division as challenging themes. "Challenging theme attack consultation rooms" were held in each division with the participation of the people who registered the challenging themes, their supervisors, the SQC promotion committee chair, and promotion committee members to promote the theme.

  This initiative was put into practice for the purpose of improving conventional guidance by supervisors that tended to make SQC utilization flat and focused too much on presentations. This meeting system is extremely practical in that people sit together in a ring to discuss themes using drawings and phenomena at planning stage. The discussion proceeds through processes such as identifying issues to setting objectives to the method of "climbing the mountain" to solve the problem.

  These meetings are held by the people who registered the themes and their supervisors a number of times in the PDCA cycle stage; it is extremely effective in follow-up and contributes a great deal to practical outcomes.

(4) Summary of practical outcomes

  As an example of the actual outcomes seen in staff SQC promotion, Fig. 6 shows the results of undertaking challenging themes. Fig. 6, diagram (1) shows the progress of extensive QCD research activity in each division. In addition, Fig. 6, diagram (2), which is divided by type of technology for which SQC is utilized, shows that themes were being undertaken that had a direct bearing on actual work in each division.

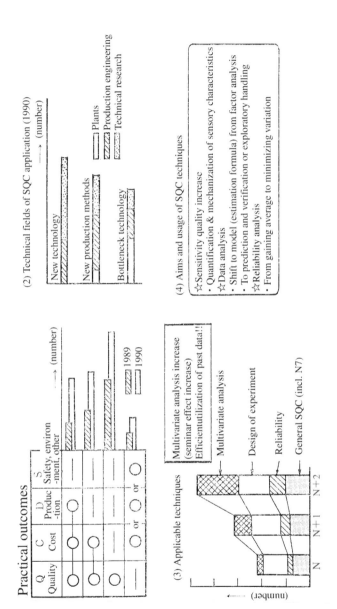

Fig. 6. The result of undertaking challenging themes

Furthermore, Fig. 6, diagram (3) shows the widespread increase in the utilization of multivariate analysis and other SQC techniques, reflecting the effect of stratified SQC training. In addition, looking at the objectives of SQC techniques and the way in which they are used as indicated in part (4) of Fig. 6 (although quantitative expressions are abbreviated), it is clear that technical staff began to utilize SQC aggressively to combat critical technical issues in order to skillfully clear hurdles that they once could not overcome.

In Fig. 5, described above, the increase in the number of extra-company presentations and the concrete examples of results reflect the status of undertaking the challenging themes. This is a sign that SQC utilization is expanding from downstream (plant divisions) to mid-stream (production engineering divisions) and upstream (research and development, design, and evaluation divisions). It also signifies that young engineers are not utilizing SQC techniques simply for the sake of SQC, but that they are beginning to use SQC well in timely QCD research activities, indicating the beginning of growth of young leaves.

References

[1] K. Amasaka,(1998), "Application of Classification and Related Methods to the SQC Renaissance in Toyota Motor", Hayashi,C et al.(Eds.),*Data Science ,Classification and Related Methods,* 684-695, *Springer.*

[2] K. Amasaka,(1989), "TQC at Toyota, Actual State of Quality Control Activities in Japan", *Union of Japanese Scientist and Engineering,The 19th Quality Control Study Team of Europe,* 39, 107-112.

[3] Toyota-wide SQC Seminar: Report of the Design Evaluation/Process Analysis/Process Control Working Groups (limited distribution version) (In Japanese).

[4] K. Amasaka,(1992),"Development of SQC Renaissance in Toyota Group", (In Japanese), *Japanese Standards Association,* 45(3), 64-68.

[5] K. Amasaka,(1995), "A Construction of SQC Intelligence System for Quick Registration and Retrieval Library,-A Visualized SQC Report for Technical Wealth-", Stochastic Modeling in Innovative Manufacturing, Anthony H Christer et al.(Eds.), *Lecture Notes in Economics and Mathematical Systems,* 445, 318-336, *Springer .*

[6] Toyota Motor Corp.,(1993), "Feature: SQC at Toyota", K.Amasaka(Eds), Toyota Technical Review, 43.

[7] M. Kamio and K. Amasaka(1992), "Collection of Activities Example Using SQC Method to Improve Engineering Technologies", *Japanese Standards Association, Nagoya QC Research Group.*

Chapter 7: Management SQC

New SQC Education for Managers' Performance

A Proposal of the New SQC Internal Education
for Management

Authors propose a new Science SQC as a demonstrative-scientific method which enables the principle of TQM to improve. They grasp the importance of Management SQC which is one of the core methods for promoting systematic and organizational development of the Science SQC and study its effectiveness. To propagate and establish the Management SQC, this report discusses the effect of the Management SQC through the development of the New SQC education implemented by Toyota Motor Corporation in recent years as a useful concept for carrying out Manager's job.

Keywords; "Science SQC", "Management SQC", "Management SQC Seminar", "TQM-S".

1. Introduction

Authors propose a new "Science SQC" as a demonstrative-scientific method which improves the principle of TQM.[1][2][3] They grasp the importance of "Management SQC" which is one of the core methods for promoting systematic and organizational development of the "Science SQC" and are further studying its effectiveness. [4][8][9][10]

To propagate and establish "Management SQC, this report demonstrates the effect of "Management SQC" through practical use and development of the demonstratively new SQC education called the "Management SQC Seminar" planned and implemented by Toyota Motor Corporation and participated by other Toyota group companies in recent years as a useful concept for carrying out managers' jobs.

2. The Development of "Science SQC" to Improve the Principle TQM[5]

To manufacture attractive merchandise to the satisfaction of customers, it is important to raise the quality of job and prevent any problems from arising under the optimal conditions. This is what the authors call TQM activities. To effectively enhance TMS, TDS and TPS which are the principal management core technologies that constitute the principle of TQM, it is necessary to develop the "Science SQC" that they have proposed as a new SQC for manufacturing industries as shown in Fig. 1.

And, here, it is important that all departments from planning, developing, design to production and sales should grasp the fundamentals of manufacturing and improve the quality of job as TQM activities "TQM-S" as shown in Fig. 2. at each stage of the business process and improve the level of engineering through the layers of demonstrative studies. [22][23][24]

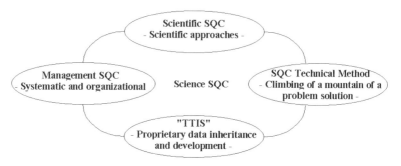

Fig. 1. Schematic Drawing of Science SQC [1]

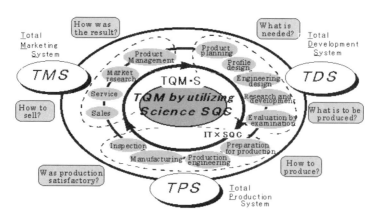

Fig. 2. TQM Activities "TQM-S" in Toyota [5]

3. Necessity of "Management SQC" [6]

The main objectives of "Science SQC" is to find a scientific solution for a gap generated between the theories (principles and laws) and reality (events). Actually, it is to find a solution that can be generalized (hereinafter referred to as general solution) by clarifying six gaps generated between the theory, calculation, test and reality as shown in Fig. 3. In other words, it is necessary to expressly describe tacit knowledge of business process of every department to make each portion of an organization better informed on others.

Therefore, the "Management SQC" is a core engineering to manage the development of "Science SQC" systematically and/or organizationally and contribute to improving quality of business process that involves decision-making on the part of managers. Its representative methodology includes the "total task management team activities" that constitute the key to the "Management SQC." [4]

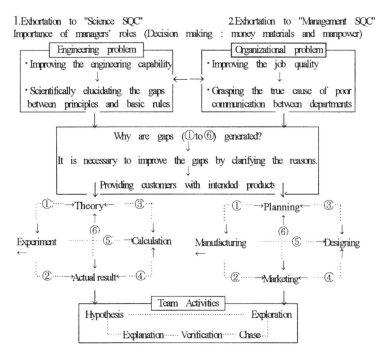

Fig. 3. Schematic Drawing of "Management SQC" for "Science SQC "Development [4]

4. Recommendation of "New SQC Education" Useful for Managers' Performance

It is considered important as a new TQM activity to vitalize both people and organization to propagate and establish the "Management SQC Seminar"[12], consisting of the following components indispensable as new SQC education curriculums, so as to help managers by improving their performance and problem solving capabilities.

4.1. Approaches of "Scientific SQC" [6]

Successful approach to SQC by the manufacturing industry lies in carrying on excellent QCD study activities. There, managers manage engineering staff. They solve important engineering problems by applying deductive method. In addition, they apply SQC to the scientific and inductive approaches to the problems by utilizing their insight to the fullest extent. The ultimate goal of the "New SQC Education" lies in acquiring the "Scientific SQC" approach that will improve managers' decision-making process quality as indicated in Table 1.

Table. 1 Objectives of "New SQC Education" for Managers

```
┌─────────────────────────────────────────────────────────────────────────┐
│ Decision-making to the reasoning of managers is important                 │
│ *   Intuition (Subjectivity + Objectivity  →  Perception)                 │
│     (Insight)                                                             │
│                                                                           │
│          ↑                                                                │
│                                                                           │
│ *   Correct judgment based on the facts                                   │
│     (Inductive method)                                                    │
│                                                                           │
│          ↑                                                                │
│          │                    ┌─ Physical and chemical engineering capability │
│ *   Proprietary technique     │                                          │
│     (Deductive method) ───────┼─ Knowledge                               │
│                               └─ Empiricism (technique)  ←  Experience    │
└─────────────────────────────────────────────────────────────────────────┘
```

4-.2. Effectiveness of "SQC Techinical Methods" [6][7]

Practice of "SQC Technical Methods", a quick problem-solving method, where the SQC too heavily dependent on analysis is withdrawn and replaced with effective combination of such steps as problem construction, task setting, examination of solution process (test and analysis design), verification, N7 to multivariate analysis, experimental design and so forth.

4-.3. Application of Integrated SQC Network "TTIS" [8][9]

Understanding and practicing a possible quick and proper climbing of a mountain of problem solution from the middle of the path by applying the proprietary SQC application case retrieval system (TSIS), SQC personal computer software (TPOS), SQC practical application manual (TSML) and the technical report (TIRS).

4-.4. Importance of "Management SQC" [6]

To solve a deep-rooted engineering problem, the total task team activities as seen in the "Total QA Network Activities"[4][5] covering the specialty engineering techniques of all the department will be effective. They key points here are understanding the importance of expressly stating implicit knowledge of business process of every department and realizing its contribution to the TQM activities as the " Management SQC." [13]

5. Opening of New SQC Course For Toyota Managers [12]

5. 1. Objectives of the Course

Managers have to grapple with the today's management problems to develop "Science SQC" to improve the principle of Toyota's TQM activities as shown in Fig. 2. Upon solving these problems, it is important for them to scientifically verify the correctness of their own thinking (theory and logic).

A "new idea" is often born from foggy problems through tests and verifications (new facts R new knowledge). [14][15] This course aims at teaching managers demonstratively the management process for general solution through the application of their proprietary techniques (empirical technique and rules) and the practice of the "Science SQC."

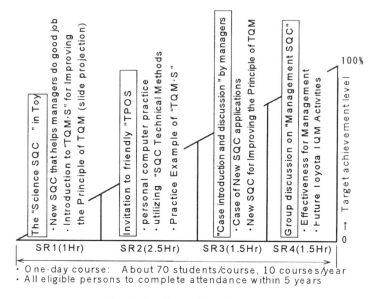

Fig. 4. Outline of Curriculum

Fig. 4 indicates the example of curriculum for a "New SQC Course for Managers" (one-day course). Particularly, managers learn mainly the process for "Developing Objective Information from Subjective Information" and "Developing Subjective Information from Objective Information" as the "Science SQC" that will contribute to the managers' decision-making. After attending the course, managers will take the initiative in managing the QCD study activities to improve the principle of business cycle in Fig. 2.

5. 2. Outline of the Curriculum

(1) SQC Useful for Managers' Good Performance

The students (managers) will start by dropping their old images on SQC. They will understand that SQC is an essential demonstrative and scientific method for solving not only reactive problems but proactive engineering problems. They will learn here the transition from the "SQC Renaissance" to the "Science SQC" and the effects of SQC contribution to the Toyota TQM activities[2][16] in a lecture type course.

During the lecture, they will be introduced to the verification cases[17][18][19] of "Management SQC" where SQC is scientifically applied and promoted for systematic and organizational development. They will acquire new way of thinking and approaches to improving the quality of management process.

Moreover, a slide on the "Science SQC"[11] will be projected to introduce to the students a new SQC that will contribute to the TQM activities. This will help the managers understand the effect of the latest "Friendly SQC"[20] that has been completely transformed from the SQC as advocated by part of specialists who are too heavily dependent on analysis.

(2) Practice of Friendly "TPOS" SQC Personal Computer Software

To support a quick and efficient job performance by freeing managers from tedious manual calculation, the SQC personal computer software is an essential tool. In this course, a program based on a scenario about the problem solution process is provided. The students learn how to use the original SQC personal computer software "Friendly TPOS"[20] so as to obtain new knowledge according to the thinking process of persons to use it. For the student managers to manage the engineering staff and carry on a smooth "total task management team" activities[17][18], each student is provided with a personal computer and they will touch TPOS in a practical training form[22].

Here, they do not learn individual methods one by one. Instead, they proceed with the practical course along the manager's business process by using the actual cases where the "SQC technical methods" are used. Fig. 5 shows the practical study menu of reactive problems. The students will understand that they can reach the top of the problem-solving mountain efficiently by effectively combining the SQC methods in each step of their business process.

During the course, several SQC special staff provide the students with powerful support to help them operate the personal computer without problem. Since "TPOS" is developed on the friendly concept, students who have never operated a personal computer can master the operation before the end of the course. They will take the TPOS software with them back to respective job-sites for actual use. Moreover, since "TPOS" in cassette form allows free combination of various SQC method libraries, it is applicable not only to reactive problems but to the application and use of various "SQC technical methods." This is the biggest advantage of TPOS.

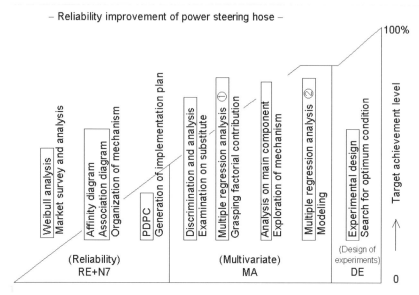

Fig. 5. Example of Mountain Climbing of Problem Solution in the Practice of
Personal Computer [21] (Application of SQC Technical Methods)

(3) "Case Introduction and Discussion" by Managers

Here, managers who actually took part in a problem-solving take initiative in introducing the cases of SQC application that practices the "Management SQC."[6] Two to three cases related to TMS (Toyota Marketing System), TDS (Toyota Development System) and TPS (Toyota Production System) are selected by field. The system is updated each time with the latest cases for implementation. For example, a manager introduces the cases mainly on the SQC application that contributes to the improvement of business process quality in the total task management team activities.

He discusses on the conception on the work and how to proceed with it from the management engineering point of view with other students. They will be strongly motivated and will attempt to develop the management SQC in their own departments. The case introducing manager will reflect the opinions expressed by the managers of respective departments from the multiple points of view on the future business process. This introduction and discussion offer an opportunity for information exchange among the men for mutual education.

(4) Group Discussion

Group discussion is held by each team under a separate subject on how to understand the Toyota's TQM activities and how to develop the "Management SQC" as a manager. Managers from respective departments that usually have no opportunity to meet with each other exchange opinion on the "Future TQM Activities of Toyota and Contribution of SQC" [4][5][6] designed to improve the business process quality and vitalize organization by exploiting the human resources to the fullest extent. This turns out both tangible and intangible effects.

5. 3. Results of Implementation

Thus far a total of about 3,000 persons (including those from Toyota group, 60 times/1996-2000) attended the course. The course is favorably evaluated by the students who say that the course helped them wipe out old images about SQC and that it provides them with an opportunity for understanding the new SQC concept. For example, some managers say, "Thanks to the course, I can now guide my staff with self-confidence" or "I am self-confident to do it myself" and so on. The course evoked new interest to SQC among the company employees to achieve expected objectives. A demonstrative study example where managers after receiving the "Management SQC Seminar" has promoted and practically provided guidance on "Management SQC".

6. Demonstrative Research by Effect of the New SQC Internal Education "Management SQC Seminar"

Table 2 [10] summarizes a part of approximately 200 studies announced to the academic associations, including Union of Japanese Scientists and Engineers (JUSE) and the Japanese Society for Quality Control. These studies were promoted in Toyota Motor Corporation during the period from SQC Renaissance and Science SQC (started in 1988). The hatched items are the practical studies of Management SQC initiated by directors and managers. The following cases are some examples of experimental studies promoted by the directors and managers who completed the recent Management SQC Seminar of Toyota Motor

Corporation.

(1)Demonstrative cases of "Science SQC" application to solving actualized engineering problems of importance include estimation of the lift characteristic of a vehicle jointly using CAE (neural network) and SQC[14] Example of Quality Assurance Activity for Brake Pad by utilizing "Management SQC"[15], and research on construction of the Toyota Development System (TDS)[30]. Another application is in "Inline and Online SQC" used in research for constructing Toyota's new production system (TPS) including the improvement of reliability and maintainability of equipment in the production plant[31].

(2)Demonstrative cases of "Science SQC" application to solving foreseeable and/or latent engineering problems include research on strategic patents[32] for enhancing intellectual assets and the development of "Design SQC" to assist merchandising planning as seen in research on profile design[33]. Moreover, in the implementation of "Marketing SQC", "Science SQC" is applied to marketing activities as a new application of demonstrative research. There, "Flyer" effect[34] and "Innovation for creating bonds with customers[35] which are useful for sales department activities represent an important foundation of the Toyota Marketing System (TMS).

7. Conclusion

(1)We have demonstrated the necessity of "Management SQC" as a core method for systematically and organically managing "Science SQC" that enhances the base of TQM.
(2)We have also indicated the importance of the "Management SQC Seminar" as a new SQC training course useful for managers in propagating and establishing "Management SQC" together with the required curriculum components.
(3)To be more definite, we have demonstratively verified the effect of SQC education through opening and practical development of the "new SQC Seminar" for Toyota managers.

Table 2. Changes in the result of the research case of Toyota (1988-2000)

References

[1] K. Amasaka et al.(1996), The Promotion of "Science SQC" in Toyota, *Proceedings of the International Conference on Quality, Yokohama, Japan,* 565-570.

[2] K. Amasaka, (1997), The Development of New SQC for Improving the Principle of TQM in Toyota, *Proceedings of the 11th Asia Quality Symposium, Taiwan,* 429-434.

[3] K. Amasaka, (1998), A Study of "Science SQC" for Improving the Principle of TQM in Toyota, *Informs 98 Seattle National Meeting, Partnering for Global Technology Management, Merger of Operations Research Society of America (ORSA) and The Institute of Management Sciences (TIMS), Washington, USA.*

[4] K. Amasaka, (1999), A Study of "Science SQC" by Utilizing "Management SQC" - A Demonstrative Study on A New SQC Concept and Procedure in the Manufacturing Industry -, *International Journal of Production Economics,* 60-61, 591-598.

[5] K. Amasaka and S. Osaki, (1999), The Promotion of New SQC Internal Education in Toyota Motor - A Proposal of "Science SQC" for Improving the Principle of TQM -, *The European Journal of Engineering Education on Maintenance,* 24(3), 259-276.

[6] K. Amasaka, (1998), Application of Classification and Related Method to the SQC Renaissance in Toyota Motor, Data Science, *Classification and Related Methods,* 684-695, *Springer.*

[7] K. Amasaka, (2000), A Demonstrative Study of A New SQC Concept and Procedure in the Manufacturing Industry - Establishment of A New Technical Method for Conducting Scientific SQC -, *An International Journal of Mathematical & Computer Modeling,* 31(10-12), 1-10.

[8] K. Amasaka et al., (1998), A New Principle of "Science SQC" for TQM Activities - TQM-S in Toyota-, (in Japanese), *JSQC (Journal of the Japanese Society for Quality Control) The 28th Annual Technical Conference,* 97-100.

[9] K. Amasaka et al., (1997), "The Development of New SQC for Improving the Principle of TQM, -From "SQC Renaissance" to "Science SQC"-", (in Japanese), *Journal of the Japanese Society for Quality Control, The 55th Technical Conference,* 13-16.

[10] K. Amasaka et al., (1996), A Study of Effectiveness of SQC for Management, (in Japanese), *The 53rd Technical Conference, Journal of the Japanese Society for Quality Control,* 85-88.

[11] K. Amasaka, (1995), A Construction of SQC Intelligence System for Quick Registration and Retrieval Library, - A Visualized SQC Report for Technical Wealth -, *Lecture Notes in Economics and Mathematical Systems*, 445, 318-336, *Springer.*

[12] K. Amasaka and T. Kosugi, (1997), A proposal of the new SQC Internal Education for Management, - The Development of "Science SQC" for Improving the Principle of TQM-, (in Japanese), *Journal of the Japanese Society for Quality Control, The 27th Annual Technical Conference,* 19-22.

[13] K. Amasaka and T. Kosugi, (1991), Application and Effects of Multivariate Analysis in TOYOTA, (in Japanese), *The Behavior Metric Society of Japan, The 19th Annual Conference,* 178-183.

[14] K. Amasaka and K. Maki, (1991), Application of Multivariate Analysis for the Attraction of Manufacturing Vehicles, (in Japanese), *The Behavior Metric Society of Japan, the 19th Annual Conference,* 190-195.

[15] K. Amasaka, (1993), SQC development and Effect at Toyota, (in Japanese), *Quality, Journal of the Japanese Society for Quality Control,* 23(4), 47-58.

[16] K. Amasaka et al. (1994), Consideration of effieientical counter measure method for Foundry, -Adaptability of defects control to Casting Iron Cylinder Block-, (in Japanese), *Journal of the Japanese Society for Quality Control, The 47th Technical Conference,* 60-65.

[17] K. Amasaka et al., (1996), A Study on Improving Disk Brake Pad Quality to Reduce Squeal, (in Japanese), *Journal of the Japanese Society for Quality Control, The 53rd Technical Conference,* 89-92.

[18] K. Amasaka et al., (1997), The Development of "Total QA Network" by utilizing "Management SQC", -Example of Quality Assurance Activity for Brake Pad-, (in Japanese), *Journal of Japanese Society for Quality Control, The 55th Technical Conference,* 17-20.

[19] K. Amasaka, (1994), SQC at Toyota , *The 29th Annual Conference of All Toyota's TQM Activities for Top Management.*

[20] K. Amasaka et al., (1995), "Aiming at the Statical Package using in the Job Process", (in Japanese), *Journal of the Japanese Society for Quality Control, The 55th Annual Technical Conference,* 3-6.

[21] M. Ouchi, (1994), Factorial Analysis on the Power Steering Hose Life Improvement, *Proceedings of the 44th NIKKAGIREN Managers and Staff Quality Control Convention,* 211-216.

[22] T. Takaoka and K. Amasaka, (1991), Derivation of Statistical Equation for Fuel Consumption in S. I. Engineers, (in Japanese), *QUALITY, Journal of the*

Japanese Society for Quality Control, 21(1), 64-69.

[23] K. Kusune, K. Amasaka et al., (1992), The Statistical Analysis of the Springback for Stamping Parts with Longitudinal Curvature,(in Japanese), *QUALITY, Journal of the Japanese Society of Quality Control,* 22(4), 24-30.

[24] K. Amasaka et al., (1993), A Study of Quality Assurance to Protect Plating Pants from Corrosion by SQC, - Improvement of Grenading Roughness for Rod Piston by Centerless Grinding-, (in Japanese), *QUALITY, Journal of the Japanese Society for Quality Control,* 23(2), 90-98.

[25] K. Amasaka, et al. (1997), The development of working condition taking the lead an epoch (#1)(#2), (in Japanese), *Journal of the Japanese Society for Quality Control, The 57th Technical Conference,* 53-60.

[26] Toyota Motor Corporation and Toyota Motor Kyushu, (1994), Development of A New Vehicle Assembly Line, (in Japanese), *Business Report on Receiving Okochi Prize for 1993 (40th),* 377-381.

[27] K. Fukumoto et al., (1997), Studies on the Ease of Use of Auxiliary Devices in the Car Assembly Process, *Journal of the Japan Society of Human Engineering,* 33, Special Edition, 148-149.

[28] O. Nishikaze, et al., (1995), Adaptation and Distortion (Wear and Repair), *The Study on Industrial Stress,* 3(1), 55-64.

[29] K. Amasaka et al., (1996), A Study of Estimating Coefficients of Lift at Vehicle, - Using Neural Network and Multivariate Analysis Method Together-, (in Japanese), *The Institute of Systems Control and Information Engineers,* 9(5), 229-235.

[30] K. Amasaka et al., (1998), A Proposal "TDS-D" by utilizing "Science SQC" - An Improving design quality for drive-train components-, (in Japanese), *Journal of the Japanese Society for Quality Control, The 60th Technical Conference,* 29-32.

[31] K. Amasaka and H. Sakai, (1998), Availability and Reliability Information Administration System "ARIM-BL" by methodology in "Inline-Online SQC", *International Journal of Reliability and Safety Engineering,* 5(1), 55-63.

[32] K. Amasaka, et al., (1996), An Investigation of engineers Recognition and Feelings about Good Patens by New SQC Method, (in Japanese), *Journal of the Japanese Society of Quality Control, The 52th Technical Conference,* 17-24.

[33] K. Amasaka et al., (1999), Studies on "Design SQC" with the Application of "Science SQC" - Improvement of Business Process Method for Automotive Profile Design -. *Japanese Journal of Sensory Evaluations,* 3(1), 21-29.

[34] K. Amasaka et al., (1997), A Factor Analysis of "CHIRASHI" Advertising

Effectiveness - The Development of "Marketing SQC" for Dealers' Sales Activities -, (in Japanese), *Journal of the Japanese Society for Quality Control, The 27th Annual Technical Conference,* 43-46.

[35] K. Amasaka et al., (1998), The development of "Marketing SQC" for Dealers' sales operating system, -for the bond between Customers and dealers-, (in Japanese), *Journal of the Japanese Society for Quality Control, The 58th Technical Conference,* 155-158.

Chapter 8: TSIS-QR System

A Visualized SQC Report for Technical Wealth:

A Construction of SQC Intelligence System for Quick Registration and Retrieval Library

Various SQC practical reports are expected to contribute to accumulating technical wealth for handing down and developing technology, and to be used as guidelines. Excellent SQC practical reports have common formulas for solving problems. Therefore, this paper proposes to systematize a visualized SQC intelligence system for the quick registration and retrieval library, by summarizing the flowcharts on solving problems of SQC practical reports and by studying both engineering technologies and SQC methods.

Keywords: SQC intelligence system, Registration, Retrieval, Technical Wealth, TSIS-QR.

1. Introduction

A wide variety of SQC(Statistical Quality Control) practical reports should be utilized as guidelines that can contribute to building up of wealth of engineering technologies as well as support for handing down and developing engineering technologies. We have taken notice of the fact that there are common formulas among excellent SQC practical reports in terms of how a technical engineer proceeds with his work and how he approaches problems.

Based on this point, we have summarized the flow of processes found in SQC practical reports and considerations from both aspects of engineering technologies and SQC methods to develop a simple flowchart developed from the three viewpoints of management, proprietary technologies and SQC so that the flow of work can be visualized. Furthermore, we have constructed TSIS-QR(*1) by systematizing the reports into "a registration and retrieval system for outlined review of SQC practical reports that allows understanding at a glance."

This is the first attempt to compile tried formulas of practices as a manual for engineers. It provides quick registration and retrieval as well as superior approaches based on practical achievements. This study discusses the aim of the efforts to construct this system that has started operation as common asset, aiming to improve the quality of engineers' and managers' work, while introducing its outlined applications.

*1 Total SQC Intelligence System for Quick Registration and Retrieval Library

2. SQC that Improves the Quality of Work

In order to deliver attractive products to users, it is required to maintain timely QCDS(*2) studies. Meanwhile, engineers' and managers' performance of high quality and competent decision making is the basis of managerial techniques. The TQM(Total Quality Management) activities supporting managerial techniques are based on a software technology useful for continuous realization of the corporate objectives. SQC, as the core of such technology is given the following "position." See, e.g., [1-4].
(1) SQC as a scientific methodology that allows visual understanding of the know-how, processes and achievements of engineers' and managers' deductive work methods as well as a common language for decision making.
(2) SQC as an inductive problem-solving methodology to probe "the gap between theory and practice" as well as practical science(behavioral science).
(3) SQC as an approaching means to grasp essential problems of technologies such as probing and verifying of complex relations of cause and effect or creation of models for prediction and control, whether in a proactive or reactive field. Technologies are constantly being accumulated and developed. The expectation for and the roles of SQC are increasing as engineers and managers demand to improve the quality of their work so that they can catch up with the advancement of technology.

*2 Quality, Cost, Delivery and Amount, Safety and Satisfaction

3. History behind Construction of TSIS-QR

3.1. In Coordination with "the SQC Renaissance"

In recent years, Toyota has been advocating "the SQC Renaissance" and promoting company-wide SQC activities as shown in Fig. 1 based on the renewed recognition of the importance of SQC (see, for example, [5-7]). The aim of this campaign is to capitalize on the SQC techniques through the entire organization including all the departments ranging from the downstream production and inspection to the midstream production technology and to the upstream product technologies (such as planning and design, research and development, and design and evaluation), aiming to improve the quality of work.

Specifically, the objectives are that all members including the engineering staff and the management should seek to reach excellent solutions for technical problems and that they should realize practical achievement by improving the proprietary technologies through SQC practices. Another aspect of the objective is to develop the SQC promotion cycle activities in which the SQC practices result in practical and full development of SQC education that facilitates effective development of human resources, which, in turn, will be reflected in the performance of operations.

In order to make this cycle permanent, it is important to secure organized planning and management that meets the current needs to solve today's technical problems and improve the practical performance. It is timely to grasp the challenge of SQC today and implement action programs(see, e.g., [7-9]). Furthermore, the key aspect of implementation is to "construct intelligent information systems" that can hand down and develop the know-how and achievement of the SQC activities that have contributed to the building up of the technological wealth, which is as indispensable to permanent promotion of SQC.

Fig. 1. Schematic Drawing of Company-wide SQC Promotion Cycle Activities

3.2. Proposal for Library of Practical Reports on SQC

As a strategic move toward "construction of a system that facilities the accumulation of technological wealth" through application of SQC, "construction of a library of practical reports on SQC that allows understanding at a glance" has been proposed and the construction of TSIS-QR has been conceived.

The following out lines the problems and summarizes the specific problems. Then, one will clarity how the construction of TSIS-QR should be implemented and consider the likely effects by looking at the processes leading to the proposal, its objectives and the points where innovative ideas have been introduced.

[Assignment]
 "To spread and utilize superior practices of SQC"
 · Even superior SQC practices tend to be forgotten over time, leaving only temporary effect.
[Problem]
 "Why aren't even excellent SQC practices referenced for better performance of operation?"
[Phenomena]
(1) It has been often said that "there is no appropriate SQC practices reported."
 · Good examples are not known due to personnel turnover.
 · It is not clear if a report is useful or not unless carefully read.
 · Essential aspects of the example may be missing.
 · The complicated methodology of SQC does not allow sufficient time for reading SQC reports.
(2) "The compilation of SQC reports is not utilized."
 · The format does not allow easy reference.
 · An instruction guide is not provided.
 · The latest edition is not available.
(3) "Opportunities to make use of SQC reports are sometimes missed."
 · Lack of knowledge of good SQC examples in the past leads to burial of them over time. (A heap of useless treasure for nouse.)
 · Lack of knowledge of poor approaches and failures in the past leads to repetition of the same mistakes, resulting in starting from the same starting line each time. (No way to climb the mountain.)
 · An extreme example is that an excellent SQC practice performed in the last year or the year before that is known only to the people directly around the

person who attended the meeting where the report was made.
- A compilation of excellent SQC reports is simply passed around with blind stamping or left among the pile on the desk. (The contents are unknown or reading the report is post poned because the contents seem to be too full.)

[Processes leading to the proposal]

(1) "To make use of the managements' views."
- To visualize managers' experience models.
- Learning about an excellent SQC report helps grasp the attitude and the approval used in that situation.

(2) "There is a common format among reports of superior practices."
- Setting up themes (identifying and selecting problems to tackle)
- Approach (how to rotate PDCA,i.e., Plan-Do-Check-Action)
- How to tackle (technical insight and data analyzing ability)

[Objectives of the proposal and ideas for the implementation]

(Objectives)
- To create a manual for technical engineers that is a compilation of superior problem-solving methods based on practical achievement.

(Ideas)
- To make SQC reports understandable in five to ten minutes.
- To make it understandable at a glance by using just one sheet of A4 size paper per case.

[Planning on how to construct TSIS-QR]
- To make it possible to share the intellectual information obained from SQC reports as technical wealth.
- To promote intelligent communication among personnel.
- To enable quick registration and retrieval of SQC practical reports.
- To obtain the latest and best approach for problem solving.

[Expected effects of TSIS-QR]
- It should enable one to grasp the currently best possible technical approval and to start from midway, instead of the very beginning, in when having to solue a problem, so that a solution may be more likely to be reached.
- It should help to make timely and efficient technical achievement of high quality (without redundancy), enabling effective QCD studies.
- "It should serve as a guideline for handing down and developing proprietary technologies and managerial techniques."

4. Construction of TSIS-QR System

4.1. Structure of the System

The TSIS-QR system comprises of a database on a host computer, as shown in Fig. 2, so that a number of technical staff members and managers can access the database. A number of terminals are networked to the host computer, enabling quick registration and retrieval of SQC reports.

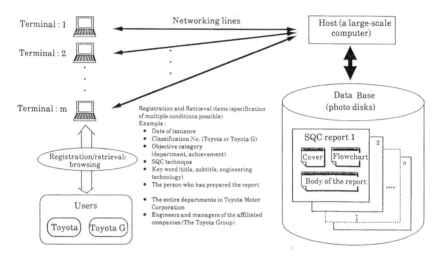

Fig. 2. Shematic Drawing of TSIS-QR System

4.2. Formats for Registration and Retrieval

4.2.1. Format for Registration

The format for registration is shown in Table 1, structured with key codes and key words. The following items are considered as required:
(Key codes)
Classification codes, objective category, department where the report is registered, etc.
(Key words)

Table 1. TSIS-QR Registration Format

TSIS-QR Registration of SQC Practical Report

① Registration No. _____ ② Number of pages _____ ③ Classification No. _____

④ Date of issuance: Day ____ Month ____ Year ____ ⑤ Objective category _____

⑥ Department where the report is registered _____

⑦ Title _____

⑧ Subtitle _____

⑨ Supplementary key word 1, SQC techniques: _____

⑩ Supplementary key word 2, engineering technologies: ____, ____, ____, ____

⑪ Reporter: _____ ⑫ Reporting department _____ ⑬ Other information _____

Table 2. Classification Code

Classification	Code	Division Contents
Major division	Z	SQC practical report
Intermediate division	T	Toyota Automobile Corporation
	G	Associated companies (the Toyota Group)
	O	Others
Subdivision	S	SQC techniques in general
	D	Experimental planning technique
	M	Multivariate analysis technique
	R	Reliability technique
	K	Sensory evaluation technique
	O	Others

Table 3. Classification of objectives

(Department Code)		(Practical Achievement Code)	
Product technology		Quality:	Q
Planning and profile design:	A	Cost:	C
Research and development:	B	Delivery and amount:	D
Design and management:	C	Safety and environment:	S
Experiment and evaluation:	D	Others:	O
Production technology and designing:	E		
Manufacture and inspection:	F		
Others:	G		

Table 4 . Retrieval Format

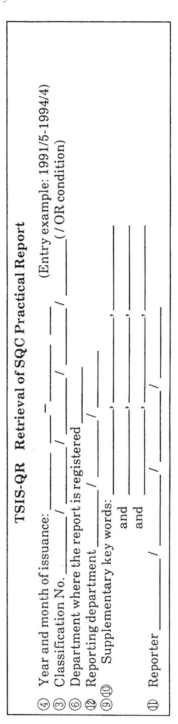

TSIS-QR Retrieval of SQC Practical Report

④ Year and month of issuance: ___ / ___ / ___ — ___ / ___ / ___   (Entry example: 1991/5-1994/4)

③ Classification No. ___ / ___ / ___ / ___   (/ OR condition)

⑥ Department where the report is registered ___

⑫ Reporting department ___ / ___ / ___

⑨⑩ Supplementary key words: ___ , ___ , ___

        and ___ / ___ , ___

        and ___ / ___ / ___

⑪ Reporter ___ / ___ / ___

Registration numbers, dates of issuance, titles, subtitles, supplementary key words (SQC techniques, terminology used for engineering technology, etc.), the names of the reporters and reporting departments, number of pages, etc.

Tables 2 and 3 summarize classification codes and objective categories that are essential components representing contents of SQC practical reports, and can be registered and indicated in several lines of characters.

4.2.2. Format for Retrieval

The format for retrieval is shown in Table 4. Display of any registered SQC report on the terminal screen is possible by entering desired retrieval conditions even if they are fragmented pieces of information. It is also possible to sequentially retrieve, browse and copy required pages of multiple SQC reports relevant for the retrieval conditions.

For that purpose, it is made possible to set a number of retrieval items so that various combinations of selected key words and codes among the registered. In particular, titles, subtitles and names of SQC techniques are divided into individual words to make them key words for retrieval for greater convenience. Furthermore, the terminals can be used for making request for copying from the photo-electronic files on the host computer as well as for retrieval.

5. Format of SQC Practical Reports for Understanding at a Glance

5.1. Composition of SQC Practical Reports

Numerous SQC practical reports stored in the TSIS-QR system shown in Fig. 2 are constructed in the "technical report format" comprising three elements; a cover, a flowchart of problem solving process and the body of the report for the purpose of accumulating technologies, and the convenience in registration and retrieval.

An A4-size sheet is allotted for the cover on page 1 and the flowchart of the problem-solving processes on page 2, respectively, summarizing the gist of the accumulated technologies and the work processes in a visualized format, which enables everybody to share information quickly.

5.2. Format of the Cover of "Technical Report"

The format of the cover includes key words and codes required for retrieval(Table 1, items 1 through 13), as shown in Table 5, to be used in handing down and developing accumulated technologies. In the text, the following items are described the relationship between the accumulation of engineering technologies and the effect of SQC is extensively covered as the gist of the technical report.

⑭ Publication and contribution
⑮ Names of computer software used for utilizing SQC techniques
⑯ Fields of technical tasks
   New technologies , new production processes, bottleneck technologies, and
   other outstanding problems
⑰ Aims of the work and its achievement
⑱ Accumulation of technologies: newly obtained knowledge and findings
⑲ Effect of SQC
⑳ PR point: Both technology and application of SQC technique
㉑ Future assignment: current undertaking to improve technology
㉒ Promotion by the secretariat: for wide propagation of the reported examples of
   SQC practical reports
㉓ Reference: technical literature, SQC practical reports, etc.

5.3. Format of "Problem-Solving Flowchart"

As shown in Table 6, the format of the "problem-solving flowchart" uses one A-4 size sheet with the following items to be used as a manual for solving problems: Technical assignments, current situation, objectives, constraints(background, factors composing the problem), flow of problem-solving processes, considerations from both aspects of technology and SQC, delivery, etc.
The points to be visualized are indicated in Fig. 3. The approach to the problem is split into its three elements of management technique, proprietary technologies and SQC analysis. The format is to split the approach into five steps -- problem selection, approach to problem, technical approach, findings, resulant out come. Pictorial symbols are employed.
In order to compress the information in an A4-size sheet, the left-hand side is

Table 5. Format of Cover of "Technical Report"

Technical Report　　　　　　　　　　Page ②

Title ⑦

	Registration No.	①
Supplementary key word 1 ⑨ SQC techniques	Classification No.	③
Subtitle ⑧	Date of issuance	④
Supplementary key word 2 ⑩ Engineering technologies	Objective category	⑤

Report of Theme for SQC Application Report Registration/Retrieval System

Reported by: ⑫ name of reporting company/department

Name of reporter ⑪

Publication and contribution ⑭　　　SQC application software ⑮

Engineering field ⑯

(1) Aim of theme and achievement ⑰

(2) Accumulation of technologies ⑱

(3) Effect of SQC ⑲

(4) Points of PR ⑳

(5) Future assignment ㉑

(6) Comments by the secretariat ㉒

(7) Reference information ㉓

Department where the report is registered ⑥ symbolic code

Reporting department ⑫ symbolic code

⑬　　Managed by: (signature)　┬ Person in charge: (signature)

Table 6. Format of "Flowchart on Problem-Solving"

used for a flowchart of the problem-solving processes. The steps clarified through SQC techniques are marked with rectangles. Newly obtained knowledge and findings are marked with rectangles with a broken line.

The right-hand side is used for considering for technologies and SQC techniques synchronized with the flow. Technically significant items are marked with, while the most significant item in terms of SQC techniques is marked with, and the most significant item in which the both technological and SQC aspects are combined is marked with.

These two flowcharts give a quick understanding of how work should be done in terms of management techniques effects of technologies and SQC techniques used in the problem-solving processes.

6. Procedures for Registration and Retrieval of SQC Practical Reports

6.1. Flow of Utilization of TSIS-QR

The procedures for registrating applications and approviing follow the flowchart shown in Fig.4 between the reporter and the SQC promotion department where

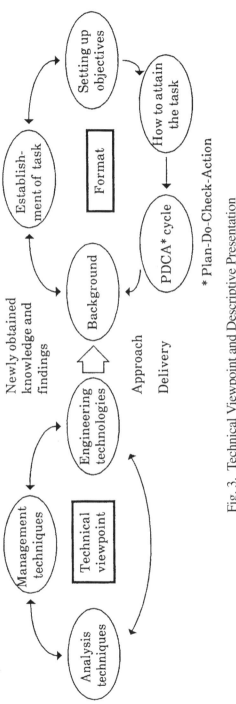

Fig. 3. Technical Viewpoint and Descriptive Presentation

* Plan-Do-Check-Action

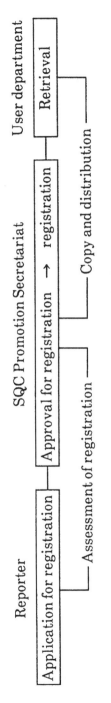

Fig. 4. Flow of "TSIS-QR"

the SQC report is to be registered. A department that desires to utilize any of the reports can then retrieve and copy from one of the TSIS-QR terminal as necessary.

6.2. Qualitative Evaluation Method for Registered SQC Practical Reports

Procedures to store good SQC practical reports in the TSIS-QR system are indispensable to SQC practical reports SQC practical reports to be registered should be processed through a qualitative evaluation system to be useful. This qualitative evaluation is made by using the format shown in Table 7. The qualitative improvement of work is compared against the general technical achievement, while the practical achievement is compared with the effect of the technological application to the product (repercussion effect).

Reports subject to registration are classified into three ranks, A, B and C. as shown in Table 7, and the respective contribution level is rated as large, medium or small. However, if no mark is given at the stage of registration application or approval, it means the qualitative level as a technological asset is low, and the report is not registered. By operating this way, the TSIS-QR system can achieve the designed target of securing practicality and reliability for reporters, SQC promotion secretariats and users.

The following introduces application examples of the TSIS-QR system operated as a "library of SQC practical reports."

7. Application Examples

7.1. Format of SQC Practical Reports

The format of SQC practical reports comprises three elements; the cover, the flowchart of problem-solving processes and the body of the report, presented in a form of a "technical report." The body of the report follows the pattern of most reports and is omitted here. Instead, the cover and flowchart will be discussed.

The technical report introduced on the format is evaluated as Rank A against the qualitative evaluation able in Table 7, and published as reference document (see, [10]), whose contents are not introduced here.

Table. 7. Qualitative Evaluation Table for SQC Practical Reports

Eval-uation Table	Effect of Appli-cation	Excellent	C	B	A
		Very Good		C	B
		Good			C
		Fair			
	Marking with a circle		Small	Medium	Large
			Generality		

Evaluation Items   *marking with a √symbol*

☐ Quality
☐ Cost
☐ Production
☐ Delivery
☐ Maintenance
☐ Safety
☐ Environment
☐ Others

The cover shown in Fig.'s 5 and 6 is the format in Section 5, structured so that all the key codes and words as well as the gist of the contents of the technical report necessary for retrieval can be obtained at a glance.

The flowchart of the problem-solving processes also uses a similar format in which the approach for problem solution, new knowledge/facts and job deadline are clearly indicated as a process flow from the top to the bottom on an A4 size sheet by paying sufficient attention to the technical viewpoint in Fig. 3.

As the problem solving flow and technological and SQC related considerations are described in synchronism, the job method can be understood at a glance. As the qualitative value of the technological wealth is clearly shown to, it is convenient for practical use.

7.2. TSIS-QR System

This section introduces application examples of the TSIS-QR system, and outlines of the operation and outputs of registration and retrieval formats in

②　全10ページ中1ページ

配付先[社外]　数部会	技　術　報　告　書		登録番号	T 9111-0008 ①

| | | | 分類記号 | Z T M ③ |

| 題目 | コンパクトDOHCシリンダーヘッド最適鋳造条件の確立 ⑦ | | 発行日付 | 1991 年 11 月 1 日 ④ |

| 手法 | 判別分析 | 3元配置 | 時系列分析 | 要因系統図⑨ | 目的区分（工数コード）の先頭5桁 | D Q C D X ⑤ |

| 狙 | 判別分析による工程内圧洩れ不良の低減 | | ⑧ | | |

| キーワード | シリンダヘッド | ダイカスト | 引け巣 | 凝固収縮欠陥⑩ | |

（海外事技術所は社外に含む）

ＳＱＣ活用事例登録・検索システムテーマ報告書
※枠内1行で簡潔に

会社名	①. トヨタ自動車　　2. トヨタ以外（会社名　　　　　　）	上 司
⑫	明知 工場　鋳造 部　技術員 ⑱・課　提供者氏名　宮本昌司 ⑪	（印）
⑭出典	「品質管理」誌　1991.5（日科技連）P 183～189	活用　トヨタ自動車
発表先	福井品質管理大会　1991.5.29（日科技連）	ソフト　TPOS-PM ⑮
⑯分 野	1. 新技術　　2. 新工法　　③. ネック技術　　④. 慢性問題	

（1）テーマの狙いと成果 ⑰

狙 い　高性能、低燃費、小型・軽量のコンパクトＤＯＨＣエンジンのシリンダヘッド粗形材の鋳造工程において、最適条件を確立し、工程内圧洩れ不良の低減を図る。

成 果　金型冷却システムの採用により、不良率が半減し（3.5％→1.8％）、年間効果額も、6,900万円／年の改善となった。

（2）技術の蓄積（得られた知見）⑱

欠陥発生メカニズムを解明し、引け巣発生要因を推定した。洗い出した7つの要因の中から下型温度、湯口温度の2つの要因に絞り込んだ。型温の状態によって、空冷や湯口ヒーターを制御するシステムを確立した。

（3）ＳＱＣの効用 ⑲

(1) 効率よく要因を絞り込むことができ、効果的対策を打つことができた。

(2) 精度の高い代用特性を見出すことができ、実験の規模を大幅に縮小できた。

（4）セールスポイント（技術とＳＱＣ手法の面）⑳

技術の面　金型温度のアンバランスによる指向性凝固の崩れで欠陥が発生するメカニズムを捉え、実験の効率と最適条件を的確に特定するため、ＤＡＳという計量的説明因子を導出した。

ＳＱＣ手法の面　凝固収縮欠陥を誘発させる要因の抽出とその効き具合を線形判別分析により、大枠でとらえ、さらに技術的洞察により、非線形判別分析へと解析技術を拡張させた。

（5）今後の課題（今、進めていること）㉑

今回の対策手法を他車種・他部門に展開できるよう、より一般化を進める。このノウハウは薄肉鋳造のアルミホイールに展開中であり、4生鋳造方案（ＣＡＥ）にフィードバックしている。

（6）事務局コメント（本事例を広く活用していただくために）㉒

製造現場に適した評価法を検討し、SQC手法を活用して、実効果が得られた。また第1回スタッフSQC全社発表会の最優秀賞（'89/1）、品質管理大会社外発表（'91/5）と活発に取り組んでいる。

事務局　（印）

（7）参考とした情報（技術情報、文献、ＳＱＣ活用事例等）㉓

技 術　軽金属学会研究委員会（1988）「アルミニウムのＤＡＳと冷却速度の測定法」

ＳＱＣ　日科技連品質管理大会（1990）「ターボ用Ｃ／Ｈのひけ巣発生の予測と制御」P241~247

配付先[社内]　数部会費金	担当部署	ＴＱＣ推進部ＳＱＣグループ ⑬	日付欄	決 裁　査 閲　認　起 案（ふりがな）
	シンボル符号	F C 0 0 0 ⑥		
合 計	依頼部署	明知工場　鋳造部　技術員室 ⑫		

Fig. 5.　Example of "Cover" for SQC Practical Reports [quoted from reference (10)]

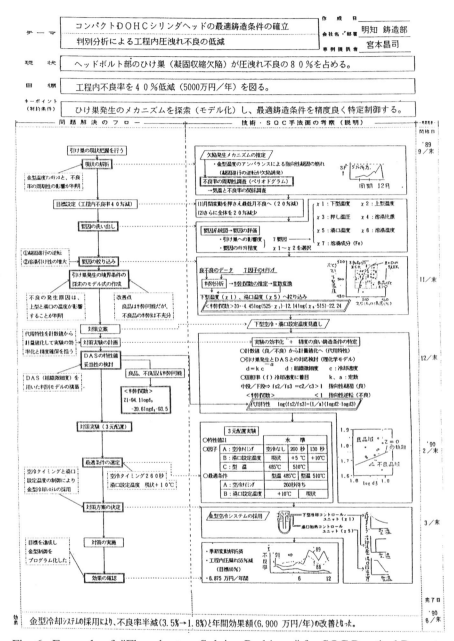

Fig. 6. Example of "Flowchart on Solving Problems" for SQC Practical Reports
[quoted from reference (10)]

particular, using examples in [6].

[Registration Format and Operation]

When a report is registered, the SQC promotion secretariat the registration by using the quality evaluation table shown in Table 7. If the report fits in either of the three ranks A, B, and C, key words and key codes are entered on the keyboard of the host computer according to the registration format (see Photo 1).The SQC practical report, which is stored on a photo disk, can be checked any time by displaying it on the large-size screen (see Photo 2).

[Retrieval Format and Operation]

A request for a retieval is made from a terminal installed in the department that desires to use the SQC system. The operating procedure is similar to that of registration. However, it is possible to specity multiple for keywords and reporter names. After browsing the large-size screen, the desired SQC reports can be retrieved and copied. (See Photo 3.)

7.3. Utilization and Effect of TSIS-QR System

The TSIS-QR system can provide users with timely and visually superior technical information. By utilizing this system, engineers and managers can retrieve registered SQC reports, and thus select relevant technical information that may be useful for their operations.

[Actual Application Example]

Individual or interactive studies on inherent technology level confirmation and problem solving processes. It allows the checking of the development of related technologies by arranging SQC application examples as a time series list. Thus, decision-making on the key points for solving the problem becomes easier through sharing of tested technical information among supervisors, subordinates and coworkers in the workshop.

[Effect of Utilization of the System]

By utilizing the TSIS-QR database, you can develop much of the solutions for a problem without unnecessary experiments or investigation. The rest of the solution can be suppouted by relevant information with a minimum amount of experiment and investigation. This enables considerable saving of cost, time and effort compared to previous methods. TSIS-QR is contributing to the quality of achievement and the reduction in development time of the QCDS activities.

TSIS-QR	New Registration		(T11)

Registration No.
  T9401-0095

No. of pages	*Classification code*	*Date of issuance*	*Objective category*
? 10	2TM	24/12/1993	# D QCDX

Title	Study on corrosion-preventive quality assurance of galvanized parts using SQC methods
Subtitle	Improvement of centerless grinding precision of piston rods
Supplementary keywords	Multivariate analysis, quantification type II, multiple regression analysis, factorial diagram. Piston rod, Grinding, surface roughness, finishing
Prepared by:	*Issuing department:*
Tsuzuki	# FC000

Photo 1. TSIS-QR Registration Format

Registration and retrieval system for SQC practical reports capitalizing on practical achievement

Photo 2. TSIS-QR Registration and Retrieval Procedure

```
TSIS-QR          Retrieval of Technical Report
   Specify retrieving conditions.
                                                 < example of entry >
   Date of issuance:     (1993                   )      1991


   Classification code   (ZTM  /        /      /      )

   Department in charge of preparation  (FC000  /          /          )

   Project code          (QCDX /        /      /      )

   Key word ( Multivariate analysis /          /          )
            and
            ( Piston    /         /          /          )
            and
            (           /         /          /          )
   Prepared by (         /         /          /          )
```

```
             Display of Retrieval Result (total one report)
   Classification: ZTM   Objective category: DQCDX    one out of total one
                                                      page
   Registration No.      Title                        Department
      Number of pages    Subtitle                     Prepared by:
                                                      FC000
   T9401-0095    Study on corrosion-preventive quality  Tsuzuki
      /p10       assurance of galvanized parts using
                 SQC methods
                 Improvement of centerless grinding precision of piston rods
```

Photo 3. TSIS-QR Retrieval Format

8. Conclusion

The TSIS-QR system started operation in late 1991 and it is contributing to qualitative improvement, higher efficiency and reduction of time of work ranging from the upstream to the downstream.

Various phenomena leading to the assignment and problems that necessitated the "proposal of a library of SQC practical reports" mentioned in section 3.2 have disappeared through the construction and operation of this system.

In recent years, an attempt has been made to propagate the application of the TSIS-QR system to our associated companies such as the Toyota Group, our suppliers, thus increasing the convenience of the application of the system.

At present, we are endeavoring to fully develop the functionality of the TSIS-QR system, by producing TSIS-RB (Reference Book, a manual), TSIS-PM (Practical Manual) and TSIS-ML (Mapping Library, a map of accumulated technologies found in SQC practical reports). We report on the establishment of a comprehensive TSIS network separately on some other chapter (See, e.g. Chapter 4).

References

[1] Amasaka, K. (1991), "What be Demanded on SQC Now-From the Front-Line of Manufacturing," (in Japanese), *Standardization and Quality Control, Japanese Standards Association,* 44(12), 79-84.

[2] Kamio, M. and Amasaka, K. (1992), "Collection of Activity Example Using SQC Method to Improve Engineering Technologies," (in Japanese), *Japanese Standards Association, NAGOYA QC Research Group.*

[3] Amasaka, K. (1993), "SQC Development and Effects at TOYOTA,"(in Japanese), *QUALITY, Journal of the Japanese Society for Quality Control,* 23(4), 47-58.

[4] TOYOTA TECHNICAL REVIEW (1993), "Special Program"SQC at TOYOTA" (in Japanese), *TOYOTA MOTOR CORPORATION, Special Number,* 43.

[5] Amasaka,K. et al.,(1991), "Re-evaluation of Present QC concept and Methodology in Autoindustry, -Deployment of SQC Renaissance in TOYOTA-," (in Japanese), *Total Quality Control, Union of Japanese Scientist and Engineers,* 42(4), 13-22.

[6] Amasaka, K. (1992), "Deployment of SQC Renaissance in TOYOTA GROUP," (in Japanese), *Standardization and Quality Control, Japanese Standards Association,* 45(5), 64-68.

[7] Amasaka, K. et al., (1991), "The Practice of SQC Education at TOYOTA -For Growing Human Resource and Practical Effort-,"(in Japanese), *QUALITY, Journal of the Japanese Society for Quality Control,* 21(1), 18-25.

[8] Amasaka, K. and Maki, K. (1992), "Application of SQC Analysis Soft in TOYOTA," (in Japanese), *QUALITY, Journal of the Japanese Society for Quality control,* 22 (2), 79-85.

[9] Amasaka, K. et al., (1994), "TOYOTA'S SQC Study System -Developing SQC Working Staff with Planning and Management Ability-," (in Japanese), *Standardization and Quality Control, Japanese Standards Association,* 47(6), 47-53.

[10] Miyamoto, S. (1991), "Establishment of Optimal Conditions for Casting Compact DOHS Cylinder Head," (in Japanese), *Total Quality Control, Union of*

Japanese Scientist and Engineers, 42(5), 183-189.

[11] Amasaka, K. and Kosugi, T. (1991), "Application and Effects of Multivariate Analysis in TOYOTA," (in Japanese), *The Journal of the Behavior Metric Society of Japan, The 19th Annual Technical Conference,* 178-183.

[12] Amasaka, K. and Maki, K. (1991), "Application of Multivariate Analysis for the Attraction of Manufacturing Vehicles," (in Japanese), *The Journal of the Behavior Metric Society of Japan, The 19th Annual Technical Conference,* 190-195.

[13] NIKKEI MEKANICAL (1994), "Special Program"SQC Renaissance in Toyota"," (in Japanese), *Nikkan Kougyo Shinbunsha,* 422, 24-35.

[14] Takaoka, T. and Amasaka, K. (1991), "Derivation of Statistical Equation for Fuel Consumption in Engines," (in Japanese), *QUALITY, Journal of the Japanese Society for Quality Control,* 21(1), 64-69.

[15] Kusune, K. et al., (1992), "The Statistical Analysis of the Springback for Stamping Parts with Longitudial Carvature," (in Japanese), *QUALITY, Journal of the Japanese Society for Quality Control,* 22(4), 24-30.

[16] Amasaka, K. et al., (1993), "Study of Quality Assurance to Protect Plating Pants from Corrosion by SQC-Improvement of Grenading Roughness for Rod Piston by Centerless Grinding-," (in Japanese), *QUALITY, Journal of the Society for Quality Control,* 23(2), 90-98.

4. Technological Quality Strategy in Toyota

Chapter 9: Products Plan Design

The Validity of the Profile Design Support Tool:

Studies on "Design SQC" with the Application of "Science SQC" -Improving of Business Process Method for Automotive Profile Design-

It is quite important for mapping up design strategies to study on "what style of vehicles would sell in the future?". To enhance the design planning quality through the application of "Science SQC", the present study aims at achieving the following: (1) To materialize vehicle images as desired by customers, (2) To analyze causal relations between the customer satisfaction assessment and the vehicle appearance review factors on the basis of the appearance style assessment, and (3) Based on the knowledge thus acquired, to understand the relevancy between vehicle images and proportion data and reflect the results to the design planning.

In the current study, the following items have been verified: We have analytically indicated the causal relationship of steps that convert the customers' sensory factor which is a direct decision-making factor for purchasing, into a concrete contour as "Design SQC" by utilizing "Science SQC". By noting that the processing of analysis itself can be part of conception and judgement of the "Designing", "SQC Technical Methods" which are a part of core technologies of "Science SQC", is applied as an analytical method. It is proposed and verified with the examples of application as the profile design support tool.

Keywords; "Science SQC", "Design SQC", Business Process, Concrete Conception Tool, Profile Design Support Tool.

1. Introduction

To develop and provide customer-oriented merchandise of attractive features, it is important for us to incorporate the feeling and words of customers into the process of new product development, that is, to implement the customer science. In recent years, Amasaka seeing the importance of SQC(Statistical Quality Control) as a behavioral science discusses its effectiveness affirmatively by proposing "Science SQC" for the quality improvement of business process of all the divisions (Amasaka et al., 1996; Amasaka, 1997, 1997,1998b; Amasaka & Osaki, 1999).

The subject is a study that constitutes the nucleus of the divisions of higher order (Amasaka, 2000). It is important for the design strategy to study on "what style of vehicles will sell in the future?". The present study proposes a methodology that will help us improve the quality of business process involved in the designing and discusses its effectiveness by quoting the examples of actual application. To be concrete, this study applies the "SQC technical methods" (Amasaka, 1998a, 2000) which are a part of core technologies of "Science SQC" to enhance the quality of the design planning job (Nunogaki et al., 1996).

The study first proceeds to materializing the vehicle images as desired by customers followed by the analysis of relevancy between the customer satisfaction assessment and the vehicle appearance review factors on the basis of the appearance style review. Moreover, on the basis of subsequently acquired knowledge, the relationship between the vehicle images and the proportion data will be grasped.

2. "Science SQC"

To provide the customer-oriented merchandise of attractive features, it is required for all the divisions concerned in the planning, development, design, manufacturing and marketing to recognize the importance of SQC as a behavioral science, understand the fundamentals of manufacture and conduct verification studies that would raise the engineering level in every step of their business processes.

The task for us today is to implement the customer science that converts customers words correctly into the engineering words. It is thus important for all the divisions to share the objective-consciousness and clearly establish the tacit knowledge on the business processing through the integrated collaboration activities. As a scientific methodology to achieve this, Fig. 1 shows the "Science SQC", a systematic and organizational application of SQC under the new concept and application method where four cores mutually help develop others.

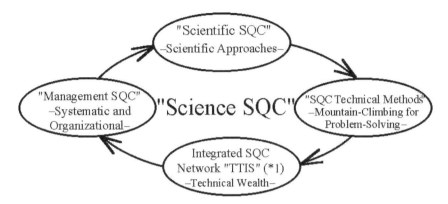

Fig. 1. Schematic Drawing of "Science SQC"

3. "Design SQC"

(1) conception and analysis in design

A concept of kansei engineering (see e.g., Shinohara et al., 1996) has been in recognition for long time. For example, Amasaka et al have developed variable data out of sensory elements such as the digitization of sensory inspection of vehicles and/or mechanization of hunch and knack work, which are being applied to actual development (Amasaka, 1972, 1976, 1983; Shimizu & Amasaka, 1975). On the other hand, examples of study (see e.g., Amasaka, 2000; Mori, 1991, 1996) are introduced where it is possible to apply a statistical method (analysis) to the development business of design, which can be interpreted as sense itself. When it comes to the application to actual business, however, there are few concrete examples of analysis worthy of introduction to a design development story for the presentation of new vehicles model.

This is attributable to the graphics that the design business (hereinafterzreferred to as "Designing") often ends up paying more importance to the end result, and that proposing good designs in the design planning or development process is based on the conception, which is not closely related with the analysis. It is apparent that the higher analysis develops, the more important the conception becomes. But the key point to the matter is how to yield a new conception, and the process of developing conception is important. In the advanced information society where identical conditions and/or data are shared, it hardly occurs to have remarkable difference in the environment.

Under the circumstances, the method for proposing a leading conception cannot be the same as before the advancement of the information society. We may be prone to use human power for what a machine can handle and may commit an error of misunderstanding that a conception has been gained. Therefore, high-quality creative activities can be carried out by determining and studying the fields assigned to the conception adapted to the times.

(2) positioning of "Design SQC"

Thus it is the objective of this study to find the guideline for establishing a method for scientifically supporting the designing so as to establish it expressly as a more creative activity from the state of tacit knowledge. It is considered that the very analysis process for establishing it as an activity would be the key to the successful conception making. It is necessary for us to create a live particular solution that catches the liking of the next generation, not the object teleology that seeks a general solution as the automotive design science. In this connection, "Science SQC" is applied to the flow of designing to actually enhance the quality of the designer's job. This is defined as "Design SQC" and applied as the activity guideline for "ADS"(*2) project.

In this report, "SQC Technical Methods" will be applied as the scientific method of mountain-climbing for the problem-solving, and by analyzing the bridge portion between analysis (research) and conception (creation of contour images), "Design SQC" will be applied as the concrete conception support tool.
(*2) Advanced Design by Utilizing "Science SQC" in Toyota

4. Guidelines and Approaches to the Study

(1) guidelines

Success of designing directly affects the sale of enterprises. Therefore, design business is established as a marketing strategy and its significance lies in the quality of the proposal. True market-in should be in proposing a desirable thing before it is desired. From Fig. 2 it is important for "Design SQC" to contribute to enhancing individual designer's proposing capability.

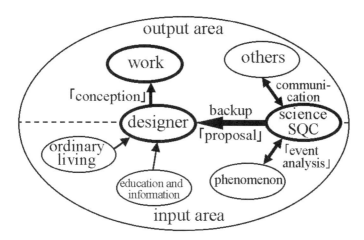

Fig. 2. Desirable relationship between "Designing" and "Design SQC"

(2) approaches

Conventionally, designing is generally developed directly to the profile design after analyzing the research itself (event analysis). In this study, bridging will be attempted to span the research-oriented analysis as the event analysis to the profile design in three steps of researches from Step 1 to 3 as Fig. 3 shows. The following state, by step, the application process of the profile design support tool that will be the guidelines for establishing a scientific methodology of "Designing" and determine the effectiveness of the "Design SQC" by quoting an application case.

Step 1 analyzes relationship between images of vehicles desirable to customers and those actually selected to search for and actualize apparent relevancy whereby a vehicle type can be specified by the desirable image.

Step 2 grasps what part of a vehicle customers observe to evaluate it. By coming down from the overall assessment, partial assessment and to detailed assessment, this report clarifies which design factor should better be given priority to satisfy customers. By thinking that true customer-in will be to propose a desirable thing before it is desired.

Step 3 classifies the vehicle proportion with the time axis (year) and high class grade (price) to grasp relevancy between the vehicle images and proportion data. This will help improve the designing process. By accumulating the improvement processes, this report intends to improve the quality of "Designing" as the dotted line indicates in Fig. 3.

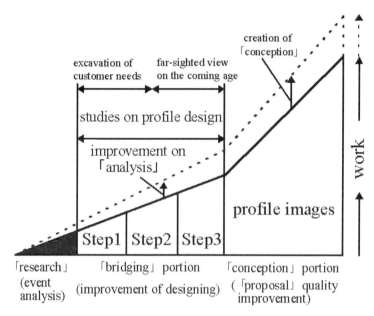

Fig. 3. Profile design support tool using "SQC Technical Methods"

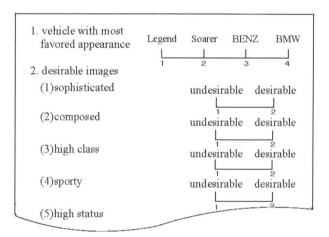

Fig. 4-2. Questionnaire on "desirable images" and the "most favored vehicle"

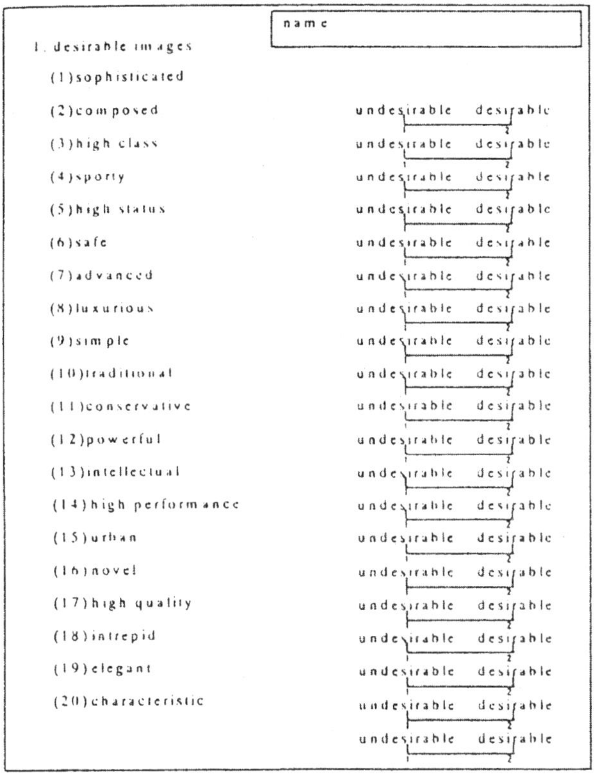

name

1. desirable images

(1) sophisticated
(2) composed
(3) high class
(4) sporty
(5) high status
(6) safe
(7) advanced
(8) luxurious
(9) simple
(10) traditional
(11) conservative
(12) powerful
(13) intellectual
(14) high performance
(15) urban
(16) novel
(17) high quality
(18) intrepid
(19) elegant
(20) characteristic

undesirable — desirable

Fig. 4-1. Questionnaire on "desirable images"

5. Application Examples

(1) Understanding correlationship between customers' desirable images of the vehicles and the most favored vehicles (Step 1)

Questionnaire was used with 157 customers (domestic panel) various in personality as shown in Fig. 4. First their desirable images are determined for the vehicles in the panel without indicating the appearance (as shown in Fig. 4-1). Then four representative vehicle models, consisting of domestic and imported models (BMW850i/1990 models, Benz300-24/1989 model, Legend Coupe/1991 model and Soarer 4.0GT/1991 model) by using photograph without vehicles' brand and name are indicated for them to answer their most favored vehicle (as shown in Fig. 4-2). Using the desirable images and the most favored vehicle data from the questionnaire, relational analysis of both data is conducted simultaneously applying the four-group discriminant analysis and the cluster analysis.

On the basis of the obtained data, panel groups belonging to BMW, Benz, Legend and Soarer are specifically extracted. Fig. 5 shows a scatter diagram of individual-score obtained as the result of the analysis using the quantification method of the third type to more clearly understand the correlationship between the desirable images and the most favored vehicle. It is understood from the figure that a panel group in favor of Soarer has desirable vehicle images such as sophistication and/or sportiness as embodied with Soarer. Since similar result has been verified with other vehicle models, it is possible to conclude that customers' liking is consistent. It was found that the customers' words can be materialized as the concrete contour of the vehicle.

(2) Method for exploring important factors customers place in vehicle appearance evaluation (Step 2)

The task taken up here is to objectify the above established theories (implicit knowledge) as qualitative, empirical rules for professional designers who plan automobile profile designs.

In established theories, Japanese users tend to lay greater stress on the front design while North American users lay stress on overall outlook including the front, side and rear views in evaluating a vehicle appearance. As far as the author know, there is no example of objectification based on investigation and analysis.

Therefore, quantitative evaluation on the sections of vehicle appearance customers are interested in will enable to advance a customer-in design strategy. Here, the same 157 customers who took part in Step 1 evaluation make an overall evaluation on the appearance of the four vehicle models. At the same time, they evaluate three vehicle appearance factors, front, side and rear views for their liking, and their cause and effect relationships are checked with multiple regression analysis as analysis I. The three appearance factors are further divided into the design balance (profile) and detailed elements (4, 9, and 5 sections respectively) for a similar study on their causal relationships as analysis II.

A preliminary cluster analysis shows that the customers can be stratified in terms of the overall liking of the vehicle appearance into a group lower in age and annual income and a group higher in age and annual income in their personalities for all four models. Fig. 6 shows an example of analytical results on a vehicle model specified to the group lower in age and annual income. Value in the figure

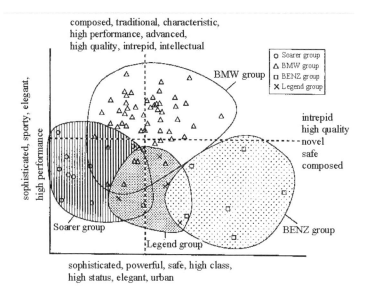

Fig. 5. Correlationship between "desirable images" and the "most favored vehicle"

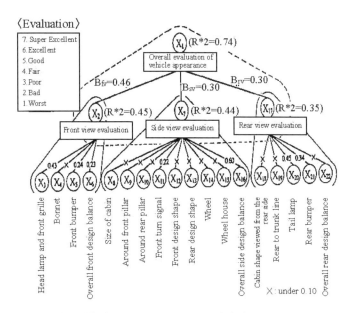

Fig. 6. Casual relationship between customor satisfaction assesment and vehicle appearance assesment factors by maltiple regression analysis

are standard partial regression coefficients B representing the degrees of influence.

In analysis I, the contributory factor adjusted for the degree of freedom ($R*2$) representing the degree of influence to the overall evaluation of vehicle appearance (X1) is 0.74, indicating a high causal relationship. The breakdown is as follows: The influence of front view (X2) is fairly high at Bfv = 0.46 while those of the side and rear views (X7 and X17) are at Bsv = 0.30 and Brv = 0.29, showing their positive influences. In analysis II, the head lamp and grille (X3) have a high degree of influence on the front view (X2) while the overall side view and design (X16) and tail lamp (X20) and rear bumper design (X21) exert much influences on the side view (X7) and rear view (X17) respectively.

In the group higher in age and annual income, though not illustrated, the influence of the front view is even higher at Bfv = 0.59 while the influence of the side and rear views (X7 and X17) are relatively low at Bsv = Brv = 0.18 in analysis I. In analysis II, the analytical results are similar to those for the group with lower age and annual income but the influence of bonnet (X4) is high on the front view (X2). It is verified that the vehicle appearance is evaluated in a wider range; for example, the influence of the line (X19) from the rear to the trunk and the design balance (X22) of the rear as a whole are high on the side view (X7). This analytical trend also applies to other three models.

A similar survey and analysis are conducted in North American market. While the weight of evaluation of the front view is generally high in Japan, it is known that the weight of evaluation of side and rear views are equivalent to that of the front view in North America. From the above, the conventional theories have analytically been tested. Through this analytical study approach, designers have understood the need for the customer-in design strategy that gives consideration to the characteristics of each country (Amasaka, 1998a,2000).

(3) Study on the customer-orientedness of profile design (Step 3)

Here, analysis will be made with the following approaches so as to estimate what would be the proportion of a vehicle that would sell in future. First, if we study why customers buy merchandise, the following two factors may be listed; They buy because they get what they pay for. They buy because the product is newer than what they have now.

By supposing that these two factors represent class feeling = price and newness = model year, relationships between these and vehicle proportion (hood ratio: hood length / overall length, trunk ratio: trunk length / overall length, cabin ratio:

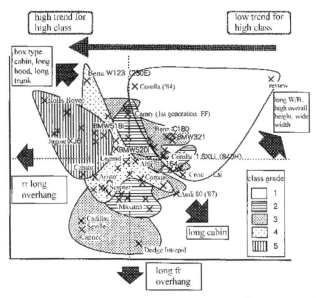

Fig. 7-1. Classification of vehicle model by the "class degree"

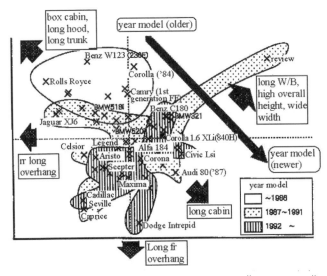

Fig. 7-2. Classification of vehicle model by the "year model"

cabin skirt length / overall length, roof ratio: roof length / overall length, front overhang ratio: front overhang length / overall length, rear overhang ratio: rear overhang length / overall length, wheel base ratio: wheel base length / overall length, roof / cabin ratio: roof length / skirt length, overall height ratio: overall height / overall length, and overall width ratio: overall width / overall length) have been analyzed. The data used for the analysis was measured with the autograph (or measurement diagram). In the case of the "Aristo" ('94 model) for example, the hood ratio is 0.26 (= hood length of 1,280mm / overall length of 4,950mm). A total of 62 vehicle models, domestic and imported (sedans) are selected to measure their proportion (ratios) and a scatter diagram (principal component score) is obtained as Fig. 7.

Fig. 7-1 and Fig. 7-2 show the result of stratified classification by the class and the model year respectively. In Fig. 7-1, the class drops toward the right of the scatter diagram and in Fig. 7-2 the model year is younger toward the right. For example, if one of these two sheets of scatter diagrams is placed on top the other, it is understood that it is hard to realize a combination of the highest class and the latest model. In other words, it can be quantitatively verified that the class and the newness are contrary to each other, which is a newly found knowledge. From this, we have been able to obtain a desirable direction for the study so as to forecast the liking of future users, not of current users, by incorporating into analysis a factor containing the concept of time axis that the customers' liking can change in terms of newness. This enables us to propose a desirable thing to customers before they want it, thereby obtaining a guideline that helps us improve the processing of design business by proposing for design process improvement.

Finally, we would like to add that the result of this study has been incorporated in the design of new "Aristo" (Motor Fan, 1997) that has been produced in a large quantity for sales.

6. Conclusions

(1) We have analytically indicated the causal relationship of steps that convert the customers' sensory factor which is a direct decision-making factor for purchasing, into a concrete contour as "Design SQC" by utilizing "Science SQC".

(2) By noting that the processing of analysis itself can be part of conception and judgement of the "Designing", "SQC Technical Methods" which are a part of core technologies of "Science SQC", is applied as an analytical method. It is proposed and verified with the examples of application as the profile design support tool.

(3) In the current study, the following items have been verified:
In step 1, customers' liking is consistent and that it is possible to materialize customers' words into contour. In step 2, customers' interest in particular elements of vehicle appearance can be stratified by their age, annual income and country. In step 3, it is possible to forecast the liking of the future users by incorporating the time axis factor of newness for the change of liking.

(4) We would like to add that the result of this study has been incorporated in a newly developed mass production model named "Aristo" intrduced to the market since 1997.
  In the future, we intend to propagate and establish design SQC by developing "Science SQC" for the enhancement and expansion of the studies on the area of bridging that leads to conception.

References

Amasaka, K. (1972). Development of Variable Data from the Inspection of Differential Carrier and Assembly for Differential Noise Control, (in Japanese) *The Second JUSE Sensory Inspection Symposium, Tokyo, Japan*, 253-263.

Amasaka, K. (1976). Differential Gear Noise Emitter and the Inspector's Sensory Characteristics, (in Japanese) *Quality Control*, 27(11), 5-12.

Amasaka, K. (1983). Mechanization of Hunch and Knack Work - Restriking of Rear Axle Shaft-, (in Japanese) *Journal of the Japanese Society for Quality Control, The 26th Annual Technical Conference, Tokyo, Japan*, 5-8.

Amasaka, K., Kosugi, T., & Ohashi, T. (1996). The Promotion of "Science SQC" in Toyota, *Proceedings of the International Conference (ICQ-96), Yokohama, Japan*, 565-570.

Amasaka, K. (1997). The Development of New SQC for Improving the Principle of TQM in Toyota - From "SQC Renaissance" to "Science SQC"-, *Proceedings of the1997 CSQC Conference and the Asia Quality Symposium, Tainan, Taiwan*, 429-434.

Amasaka, K. (1998a). Application of Classification and Related Method to the SQC Renaissance in Toyota Motor, Hayashi, C et al.(Eds.),*Data Science ,Classification and Related Methods, Springer,* 684-695.

Amasaka, K. (1998b). A Study of "Science SQC" for Improving the Principle of TQM in Toyota, *Marger of Operation Research Society of America (ORSA) and the Institute of Management Science (TIMS), Seatle, Washinton, USA.*

Amasaka, K. (1999), A Study on "Science SQC" by Utilizing "Management SQC" -A Demonstrative Study on A New SQC Concept and Procedure in the Manufacturing Industry-, *Journal of Production Economics,* 60-61.591-598.

Amasaka, K. (2000). A Demonstrative Study of A New SQC Concept and Procedure in the Manufacturing Industry -Establishment of A New Technical Method for Conducting Scientific SQC-, *An International Journal of Mathematical & Computer modeling,* 31(10-12),1-10.

Amasaka, K. & Ozaki, S. (1999). The Promotion of New SQC Internal Education in Toyota Motor -A Proposal of "Science SQC" for Improving the Principle of TQM-, *European Journal of Engineering Education (EJEE) , Research and Education in Reliability, Maintenance, Quality Control ,Risk and Safety ,*24(3), 259-276.

Motor Fan (1997), "All of New Model ARISTO", (in Japanese) *Motor Fan, A New Model Prompt Report,* Separate Volume 213, 24-30.

Mori, N. (1991). Design Engineering, -Design Engineering of Software System-, (in Japanese) *Asakura Shoten, Tokyo.*

Mori, N. (1996). Left-brain Designing, -Researching Scientific Method of Designing-, (in Japanese) *Kaibun-do, Tokyo.*

Nunogaki, N., Shibata, K., Nagaya, Ohashi, T., & Amasaka, K. (1996). A Study on customer's Direction about Designing Vehicle's Profile, (in Japanese) *Journal of the Japanese Society for Quality Control, The 26th Annual Technical Conference, Tokyo, Japan,* 23-26.

Shimizu, H., & Amasaka, K. (1975), Quality Assurance on Dead Slow Steering Effort, (in Japanese) *Quality Control,* 26(11), 42-46.

Shinohara, A., Yoshio, O., & Sakamoto, H. (1996). An Invitation of Kansei Engineering, (in Japanese) *Morikita Shuppan,* (1996).

Chapter 10: Development Design

Verification of SFC Estimation Formula:

Analysis of Factors for Specific Fuel Consumption Improvement

Recently, the need for improving the fuel economy of automotive engines has been increasing sharply. Since the fuel economy (or SFC: specific fuel consumption) depends on multiple factors that affect on another in a complicated manner, it is very difficult to estimate it in the design phase. However, its estimation in the design phase, if possible, would bring about such merits as decreasing the number of prototype engines to be made. We have, therefore, conducted a multiple regression analysis using the major specifications of Toyota engines as explanatory variables to work out a highly reliable equation for estimation and drawn up some guidelines for SFC improvement.

Keywords; Specifictual consumption, automatic engines, a multiple regression analysis, a highly reliable equation for SFC improvement

1. SFC of Engine

Figure 1 outlines how the equation for estimation has been worked out. About 30% of the fuel's energy is wasted as exhaust loss, about 30% cooling loss, and about 15% as abrasion loss, leaving only about 25% for shaft output. The methods for expressing SFC can be broadly classified into two types as summarized in Table 1. One uses whole vehicle for measurement, and the other uses the engine alone. Since the result of the former greatly varies with vehicle specifications, it is not suitable for evaluating the engine SFC.

Therefore, we used the latter for our analysis. The SFC indicating the weight of the fuel per unit work load is used as the factor representing the engine fuel economy because it can easily be handled as a function of an explanatory variable. Though SFC varies with the engine speed, charging efficiency and air-fuel ratio, these can be assumed as fixed values.

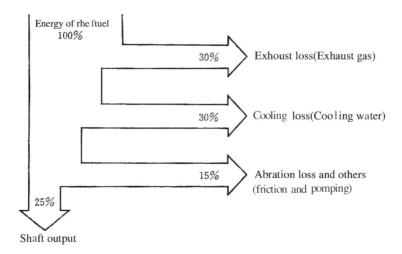

Fig. 1. The outline of the energy balance of the engine

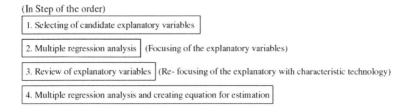

Fig. 2. The flow chart of the analytic process

Table 1. The methods for expressing SFC

Measurement form	Unit	Complement
Whole Vehiclefor measurement	km/l	How much km can this vehicle run with one liter of gasoline?
	mile/g	How much km can this vehicle run with one gallon of gasoline?
Engine alone	g/sec	How much gram of gasoline does this engine consume in one second?
	g/kw.h	How much gram of gasoline does this engine consume through the work of 1kw.h is done?

2. Multiple Regression Analysis

We have conducted SFC measurement for multiple engines having different specifications as shown in Table 2 and an analysis according to the procedure shown in Figure 2. We used TPOS-PM (Toyota TPOS, Total SQC Promotional Software for multivariate analysis) for calculations. When conducting a multiple regression analysis, we used a forward-backward selection assuming two kinds of Fin and Fout.

Table 2. The engine used for the Analysis

Number of cylinders	4~8
Displacement	1500^{cc}~4500^{cc}
Compression ratio	8.8~10.5
Bore/stroak	78.7~100^{mm}/71.5~95^{mm}

2.1. Selecting Candidate Explanatory Variables

We selected nine variables - which seem to be associated with SFC in terms of inherent technology - from engine specifications as candidate explanatory variables as listed in Table 3. A supplementary explanation of these explanatory variables is given below based on the engine sectional view shown in Figure 3.

$X(1)$: Bore (B) = Diameter of combustion chamber
$X(2)$: Stroke (St) = Stroke of piston
$X(3)$: Connecting rod ratio = Connecting rod length $l/r = 2l/St$
$X(4)$: Compression ratio (ε) = Volume α / Volume β
$X(5)$: Displacement = (Volume α - Volume β) x Number of cylinders
$X(9)$: Fuel = Type of fuel used for the test (defined variables for two types of fuel
        used)

The raw data of the above nine explanatory variables, basic statistics, and multivariate correlation chart are omitted here because of the limited space. Some technically significant physicochemical data contributed to the data analysis in the scatter diagram of (correlation between) $X(7)(= St/B)$ and SFC as shown in Figure 4.

2.2. Multiple Regression Analysis Result (1)

Table 3. Candidate explanatory variables

$X(1)$	Bore (B)
$X(2)$	Stroke (St)
$X(3)$	Connecting rod ratio
$X(4)$	Compression ratio (ε)
$X(5)$	Displacement
$X(6)$	Number of cylinders
$X(7)$	St/B
$X(8)$	Displacement/cylinder
$X(9)$	Fuel

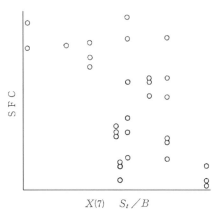

Fig. 4. The scatter diagram of *St/B* and SFC

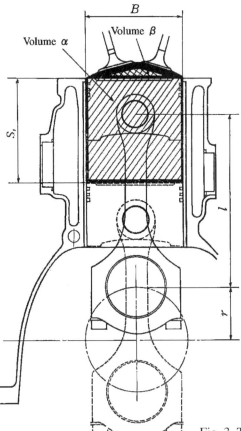

Fig. 3. The cross section of the engine

There are some stratified factors. Since the number of pieces of data ($n = 30$) is limited, we did not conduct a stratified multiple regression analysis, but we tried to know how only the main factors affects the SFC. As shown in Table 4, collected data for nine variables listed in Table 3 were input.

Table 5 shows the results of multiple regression analysis. Naturally, skewness exists in the analytical result due to normality and the correlation between explanatory variables. For example, $X(7)$(St/B) is acceptable technically but $X(9)$ (fuel), $X(3)$ (connecting rod ratio), and unselected $X(4)$(compression ratio) are underestimated. On the contrary, $X(6)$(number of cylinders) is overestimated, losing technical coherence.

Figure 5 shows the relationship between the actually measured and estimated values. A disadvantageous characteristic is also found in the regression residual as indicated by the dash line. It is obvious that simple, primary binding of four variables (linear multiple regression analysis) is deficient. Furthermore, the regression residual (σ e^2) is too large as the working-level estimation margin. As discussed above, this analysis provided us with some technical data necessary for conducting the analysis in the next step.

Table 4. Input data example (Data matrix by index-ization)

	$X 1$	$X 2$	·	·	·	·	$X 8$	$X 9$	SFC
Case 1	87.5	82.5	·	·	·	·	496	0	281
Case 2	87.5	82.0	·	·	·	·	493	1	271
·	·	·	·	·	·	·	·	·	·
·	·	·	·	·	·	·	·	·	·

Table 5. Multiple regression analysis result (1)

Multiple correlation coefficient ; 0.8712   Contribution rate(R^2) ; 0.759   Contribution rate for degree of freedom (R^{*2}) ; 0.739

Explanatory variables	Standard partial regression coefficient	Partial correlation coefficient	Standard error of standard partial regression coefficient	F-value	VIF
$X(7)$ St/B	-0.82	-0.81	0.12	47.9**	1.7
$X(6)$ Number of cylinders	-0.66	-0.60	0.17	14.4**	3.6
$X(9)$ Fuel	-0.28	-0.37	0.14	4.1	2.4
$X(3)$ Connecting rod ratio	-0.29	-0.24	0.23	2.1	6.7

Note**: Significant in 1%

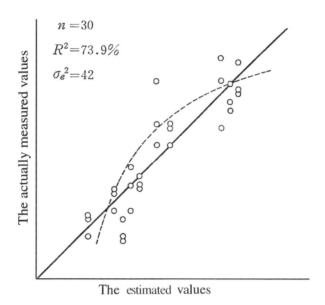

Fig. 5. The relationship between the actually measured and estimated values
(Multiple regression analysis result (1))

2.3. Review of Explanatory Variables

To work out an estimation formula which is highly reliable and technically reasonable, it is necessary to enhance the data analysis accuracy. To work out an estimation formula that is rich in universality, suitable for extrapolation, and convenient to use, we attempted conversion and selection of explanatory variables.

(1) Converted explanatory variables

1) $X(7)$ (St/B)
This is the factor that represents a heat radiation area relative to the calorific value. Increase in the heat radiation area impairs the heat efficiency and SFC. It is a known fact that the S/V ratio (surface area/volume ratio at the top dead center), or the factor representing the heat efficiency, has approximately a linear functional relationship with the SFC.

However, since it was difficult to measure the S/V ratio, we approximated the S/V as the function of B, St, and ε with equation (1) taking into account the cylinder model. Thus, we made a new explanatory variable.

$$S/V: X(10) = 2\,(\,\varepsilon - 1)/St + 4/B \qquad (1)$$

As the S/V ratio is the data representing a ratio, we hesitated as to whether it should be converted to a log value and we decided to use the S/V ratio as it was because the data has dimensions.

2) $X(3)$ (Connecting rod ratio)
This is the factor representing friction between the piston and cylinder. Since it is a dimensionless explanatory variable, we converted it to a logarithmic value (equation (2)) taking into account the addition theorem of data, thus succeeded in improvement of skewness and kurtosis and enhancement of normality.

$$\text{Connecting rod ratio: } X(11) = \log X(3) \qquad (2)$$

3) $X(4)$ (Compression ratio)
As discussed in Section 3.2, $X(4)$ was not selected as an explanatory variable because the F-value was very small. From the inherent technology's point of view, the SFC should go down with the increase in the compression ratio (a factor representing thermal efficiency). Accordingly, we made the conversion indicated by equation (3) to establish a linear functional relationship with the SFC based on thermodynamics.

$$\varepsilon\,' = 1 - 1/\varepsilon^{\,\kappa-1} \qquad (3)$$
(ε : Compression ratio, κ : Ratio of specific heat = Approx. 1.4)

As $\varepsilon\,'$ ranges from 0% to 100%, we made a logit conversion as indicated by equation (4), thus succeeded in enhancing the data analysis accuracy.

$$X(12) = -10 \log(1/\varepsilon\,' - 1) \qquad (4)$$

(2) Omitted explanatory variables

1) $X(6)$ (Number of cylinders)

Let's take at a look at the problem not discussed in Section 3.2. From the inherent technology's point of view, the plus and minus signs of the partial regression coefficient are inverted and the standard partial regression coefficient is too large. To solve this problem, we conducted a multiple regression analysis using a quantitative (defined) variable instead of a continuous quantity in vain.

$X(6)$ shows a technically significant correlation with $X(5)$ (displacement) and $X(3)$ (connecting rod ratio) and is included in other explanatory variables such as $X(8)$ (displacement/number of cylinders). Therefore, we omitted it to correct the distortion of the analysis result.

2.4. Multiple Regression Analysis and Creating Equation for Estimation

(1) Multiple regression analysis result (2)

We conducted a multiple regression analysis again using the explanatory variables reviewed in Section 3.3 as shown in Table 6. $X(7)$, $X(4)$, and $X(3)$ were omitted and $X(10)$, $X(12)$, and $X(11)$ were used for them. Before conducting the multiple regression analysis, we studied the multivariate correlation chart and basic statistics, finding that the new explanatory variables are effective and acceptable technically.

We also conducted a multiple regression analysis using a designation selection method focusing on the correlation between $X(10)$ and $X(1)$ and the correlation between the group of $X(2)$, $X(5)$, $X(12)$, and $X(8)$ and the group of $X(1)$, $X(2)$, and $X(5)$, succeeded in selecting reasonable explanatory variables.

Table 7 shows the result of analysis conducted based on the forward-backward selection method. It shows some improvements made by reviewing the explanatory variables. First, each of the selected explanatory variables $X(10)(S/V)$, $X(12)$(compression ratio), $X(11)$(connecting rod ratio), and $X(9)$(fuel) is 1% significant respectively and therefore we could obtain a reasonable result.

Furthermore, the standard partial regression coefficient and its standard error are technically reasonable. Second, the contribution rate R^2 (the contribution rate for degree of freedom R^{*2}) was improved from 75.9% (73.9%) to 81.7% (78.6%) and the regression residual (σe^2) was also improved 12%, from 42 to 37. Third, the disadvantageous characteristic of the regression remainder shown in Figure 5 is not found in Figure 6 that indicates the relationship between the actually measured values and the estimated values. Fourth, VIF (less than 10) are

Table 6. The explanatory variable reviewed

$X(10)$	S/V ratio
$X(11)$	log Connecting rod ratio
$X(12)$	$-10 \times \log\ (1/\varepsilon'-1)$
$X(9)$	Fuel

Table 7. Multiple regression analysis result (2)

Analysis of variance table

	Degree of freedom	Sum of squares	Deviation of Sum of squares	F-value
	4	4124.06	1031.01	27.8682**
	25	924.90	37.00	
Total	29	5048.96		

No.	Partial regression coefficient	Standard partial regression coefficient	Partial correlation coefficient	Standard error of partial regression coefficient	Standard error of Standard partial regression coefficient	F-value	VIF
X(10)	877.0500	2.1458	0.89759	86.15070	0.21078	103.64100**	6.0633
X(12)	-168.1490	-2.2203	-0.88166	17.99920	0.23767	87.27330**	7.7089
X(11)	-517.1410	-1.0013	-0.77941	83.13850	0.16098	38.69130**	3.5365
X(9)	-21.0553	-0.8115	-0.63774	5.08603	0.19602	17.13820**	5.2440
Constant	623.24400			55.64830			

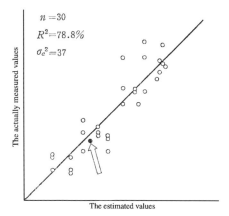

Figure 6. The relationship between the actually measured and estimated values (multiple regression analysis result 2)

completely reasonable as the important information necessary for regression diagnosis.

(2) Creating equation for estimating SFC

Using the analysis results mentioned above, we succeeded in working out equation (5) as the working-level SFC estimation formula.

$$SFC = 623 + 877X(10) - 168X(12) - 517X(11) - 21X(9) \qquad (5)$$

Scaling this equation gives equation (5)'

$$SFC' = 2.15X(10)' - 2.22X(12)' - 1.00X(11)' - 0.81X(9)' \qquad (5)'$$

3. Verification of SFC Estimation Formula

To verify our SFC we conducted a verification test on a different engine and obtained the following result (the black dot · ($\Leftarrow$) in Figure 6):

Estimated value = 277
Actually measured value = 272
Residual = 5 (1.8%)

Although n was assumed to be 1, we could verify the accuracy or reliability of the estimation formula.

4. Conclusion

A highly reliable, working-level estimation formula has been obtained to enable us to analyze the causal relations among engine specifications and the SFC using the multiple regression analysis technique. Thus, the guideline for efficient determination of engine specifications for SFC improvement has been obtained.

5. Supplements

Let's take a look at some topics not discussed above. As obvious from the data included in this paper, only a small amount of data can be obtained through one development project. However, we have obtained a variety of data from several projects carried out so far. Specifically, the Research and Development Department has data with detailed experimental conditions even through they are time-series sample data. This department has a lot of clean data unlike the downstream departments, allowing us to use it with ease.

For the stratified data used in our study, it is very important to rearrange and use the past technical information and study differentiation of the obtained data by engine types, design specifications, and experimental conditions before and after the analysis. It is necessary to determine whether there is a stratification factor according to a multi-variate correlation chart and graphed time-series data. If we can find technically reasonable differentiation information that contributes to the enhancement of analytical accuracy, we need to convert data one by one making effective use of the sequential conversion method before and after the analysis.

In addition, we effectively used the technical information obtained from regression diagnosis and various experimental data, carried out sequential transformation of some variables, and attempted to conduct a multiple regression analysis without increasing the number of explanatory variables. Details on this attempt are not given in this paper because space does not allow.

Recently, Toyota Motor Corporation announced similar studies [6]-[17] in which the scientific insight were the key to the findings. To this end, the "Science SQC" approach is promoted as a excellent technology of scientific analysis in the company.

References

[1] T,Takaoka and K,Amasaka,(1991), Derivation of Statistical Equation for Fuel Consumption in S.I. Engines, (in Japanese) *Quality*, 21(1), 64-69.
[2] H,Sono et al.,(1990), Relationship between Total Engine Loss and Major Specifications of a 4-cycle Gasoline Engine,(in Japancsc) *The 8th Internal Combustion Engine Symposium*, 375-380.
[3] K,Amasaka and K.Maki,(1992), Applications of SQC Analysis Soft in Toyota, (in Japanese) *Quality*, 22(2), 79-85.

[4] K.Amasaka et al.,(1992), The SQC use case collection that characteristic technology is extended, *Nagoya QC Research Group Edition, Japanese Standard Association Press.*

[5] K.Amasaka et al.,(2000), Science SQC-The quality reform of the business process, *Nagoya QST Research Group Edition, Japanese Standard Association Press.*

[6] S. Abe and M. Tokoro,(1989), " Research into Estimation of Engine Heat Radiation", (in Japanese) *Union of Japanese Scientists and Engineers, 13th Multivariate Analysis Symposium,* 9-15.

[7] S. Shiraya,(1990), " A Study of the Estimation Method for Sheet Blank Shape", (in Japanese) *The Japanese Society for Quality Control, 38th Technical Conference,*49-52.

[8] M.Makiguchi et al., (1991), "Analysis of Optical Distortion for Automotive Windshields", (in Japanese) *Union of Japanese Scientists and Engineers, 15th Multivariate Analysis Symposium,* 65-70.

[9] K. Yamamoto et al., (1991), "High Reliability Hybrid Circuits for Automotive Applications", (in Japanese) *Union of Japanese Scientists and Engineers, 22nd R & M Symposium,* 327-336.

[10] T. Hashimoto,(1991), " Development of Solid Lubricant Coated Piston", (in Japanese) *Quality Control, Union of Japanese Scientists and Engineers,* 42 (5), 62-66.

[11] S. Fujiwara, (1992), "Analysis of Causes for Reduction of Connecting Rod Deformation", (in Japanese) *Quality Control, Union of Japanese Scientists and Engineers,* 43(5), 281-285.

[12] S. Ado et al., (1992), "Setting Forming Contraction Rate for R-RIM Bumper Facia", (in Japanese) *Quality Control, Union of Japanese Scientists and Engineers,* 43(5), 68-74.

[13] H. Hattori et al., (1992), "Evaluation of Depth in Automotive Painting", (in Japanese) *Union of Japanese Scientists and Engineers, 22nd Sensory Inspection Symposium,* 67-72.

[14] Y. Genma, (1992), "Development of Ferrite Heat Resistant Cast Steel", (in Japanese) *Quality Control, Union of Japanese Scientists and Engineers,* 43 (5), 28-31.

[15] T. Otsuka (1992), "Analysis of Submarining Conditions in Frontal Collision", (in Japanese) *Quality Control, Union of Japanese Scientists and Engineers,* 48 (5), 236-239.

[16] K. Hatta, (1992), "Development of Large-size Thermoplastic Polyurethane

Protection Molding-", (in Japanese) *Union of Japanese Scientists and Engineers, 16th Multivariate Analysis Symposium,* 73-80.

[17] Y. Oshimura,(1992), " Qualitification of Automotive Door Closing Sound", (in Japanese) *Union of Japanese Scientists and Engineers, 22nd Sensory Inspection Symposium,* 81-86.

Chapter 11: Reliability Design

Understanding the Oil Leakage Mechanism:

Reliability of Oil Seal for Transaxle -
A Science SQC Approach

This case study deals with the reliability improvement of a transaxle unit of an automobile. This is important as the cost to rectify failures of such units is high and has an impact on customer satisfaction. We outline the approach taken by Toyota and the component manufacturer to achieve this. This involved a scientific approach to the analysis of the oil leakage mechanism through a "Dual Total Task Management Team" involving Toyota and the component manufacturer. The visualization of the oil seal leakage dynamics and the use of factor analysis based on "Science SQC" were two critical elements of this approach. This led to a better understanding of the oil leakage mechanism and changes in design to achieve higher reliability.

Keywords; Oil Seal, Transaxle unit, "Dual Total Task Management Team", "Science SQC", the oil leakage mechanism, higher reliability.

1. Introduction

It is very critical for both vehicle and parts manufacturers worldwide to improve drive train system reliability for ensuring higher customer satisfaction. The reliability of the drive train depends on the reliability of its components. One such component is the oil seal for the transaxle. Failure of this results in oil leaking out and this can have serious implications for the drive train, resulting in high repair cost and high customer dissatisfaction.

Toyota Motor Corporation purchased this component from the NOK Corporation, an external component manufacturer and developed a new approach to improve the reliability of the oil seal. Since the dynamics of the oil leak were not well understood, the starting point of the joint investigation was to get a better understanding of this. This was done through a "Dual Total Management Team" involving both Toyota and NOK using the "Science SQC" developed by Toyota. The "Science SQC" involves the "Management SQC" and the "SQC Technical Methods". Toyota and NOK formed the "TDOS-Q5" (teams Q1 to Q5) and "NDOS-Q8" (teams Q1 to Q8) teams to carry out the scientific investigation. Each team had specific goals and objectives. Three management methods, namely "Technical Management (TM)", "Production Management (PM)" and "Information Management (IM)", were adopted to develop new technologies to understand and design a better oil seal for the transaxle.

   This led to two new technologies. The first was a technology to visualize the oil seal leakage dynamic behavior in order to understand the underlying mechanisms, and the second was the use of factor analysis for improving reliability through better design. Use of the "Total QA Network" method made it possible to build quality into the process, leading to reliability improvement that reduced market claims by 90%.

   This case study discusses the approach used by Toyota to improve the reliability of the oil seal for transaxles. The outline is as follows. Section 2 deals with the sealing function of the oil seal and earlier approaches to seal design. Section 3 describes reliability improvement activities at Toyota using a cooperative approach. Section 4 deals with reliability improvement of the oil seal for quality assurance of the transaxle. Finally, in Section 5, we conclude with some comments.

2. Oil Seal

2.1. Oil Seal Function

   An oil seal, shown in Fig. 1, prevents the oil lubricant within the drive system from leaking from the drive shaft. It is comprised of a rubber lip molded onto a round metal casing. The rubber lip grips the surface of the shaft around its entire circumference, thus creating a physical oil barrier. A garter spring behind the rubber lip increases the grip of the lip on the rubber shaft. As the shaft rotates, a minute quantity of the sealed oil forms a thin lubricating film between the

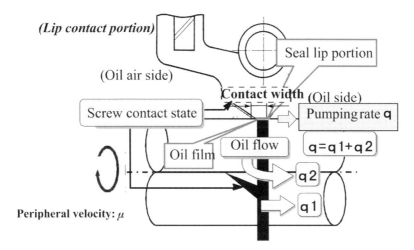

Fig. 1. Enlarged view of oil seal sealing mechanism

stationary rubber lip and the rotating shaft. This oil film prevents excessive wear of the rubber lip, and at the same time reduces frictional loss due to shaft rotation.

A properly designed rubber lip rides on this lubricating oil film. On the other hand, an excessively thick oil film will itself be a source of leakage. This condition is avoided by precise control of the lubricating oil film between the rubber lip and the rotating shaft.

Extensive experimental and theoretical investigations over the years by NOK have identified several important factors affecting the sealing function of the rubber lip. Of primary importance is the sealing ability of microscopic roughness on the rubber surface. It has been shown that these micro-asperities create shear flows within the lubricating oil film (Nakamura, 1987). The objective of the seal design (Kawahara et al., 1980) is to influence this shear flow so that the net flow of oil is towards the sealed oil side. The pumping ability of the rubber micro-asperities is designated by microscopic pump flow $q1$. To supplement $q1$, the rubber lip has also been augmented with helical ribs which function similar to a vane pump, providing macroscopic pumping $q2$ of leaked oil back to the sealed oil side (Lopez, et al., 1997, and Sato, et al., 1999). Recent efforts by NOK to maximize $q2$ have resulted in a patented rib shape (Kameike et al., 1999).

The parameters for the sealing condition of the oil film involve not only the design of the seal itself, but also external factors such as shaft surface conditions, shaft eccentricity, etc. (Hirabayashi et al., 1979). Of particular importance is the contamination of the oil by minute particles. Since these are technical issues which involve not only the seal but also the entire drive train of the vehicle, Toyota initiated a corroborative effort with NOK to improve the reliability of the oil seal.

2.2. Earlier Approaches to Seal Design

The design quality and the total quality assurance program prior to the recognition of the "transaxle oil seal leakage" problem were mostly centered on treating each part of the item separately. The technical development design staff recovered oil seal units having leakage and analyzed the cause based on proprietary techniques. Corrective measures were then incorporated into the design. Inspection of items with leakage revealed no reason for the leakage and the cause of the oil leak was labeled "unknown", which made it difficult to find a permanent solution to the leaking problem.

To develop an epoch-making quality improvement, it was necessary to study the transaxle as a whole, rather then looking at each part separately by specialists, so as to understand the mechanism leading to seal failure and the effect of the operations during the manufacture process. Two issues of importance in the manufacture of highly reliable transaxle units from a product design viewpoint were identified as follows.

(1) It is difficult to characterize theoretically the variation of macroscopic pressure distribution due to shaft eccentricity and the variation of microscopic pressure distribution under the influence of foreign matter in the oil.
(2) Design specifications relating to shaft (to mesh with oil seals), on the other hand, were decided by the vehicle manufacturer in the ranges recommended by the component manufacturer based on limited information exchange between the two parties.

For the new approach to yield the optimal design required each party to have the implicit knowledge known to the other and this was not the case. This highlighted the fact that there were some deficiencies in the total quality assurance program. In order to tackle these issues properly, a new methodology was needed and this is discussed in the next section.

The main objective of the new method was to isolate the true cause of failure.

For this reason, it was necessary to implement the "Science SQC" approach (see Amasaka, 1997 and Amasaka and Osaki, 1999) and carry out a proper study of the mechanism of the sealing performance based on a systematic approach. To achieve this objective, new task management teams comprising of members from related divisions from both the Toyota and NOK organizations were established to improve the reliability of the oil seal.

3. Reliability Improvement at Toyota: A Cooperative Team Approach

3.1. Dual Total Task Management Team

In the automotive industry (and in other general assembly industries), quality control for parts and units, optimization of adaptation technologies for assembly, and quality assurance are required in all phases of the operations (production, sales and after-sales service). Effective solution of technical problems requires the formation of teams and an understanding of the essence of problems by the teams as a whole. This allows the bundling of empirical skills of individuals distributed throughout the organization. Solution to technical problems requires harnessing the information (implicit knowledge) among related units of the organization through a cooperative team approach to generate new technologies (explicit knowledge).

If collaborative team activities by the vehicle manufacturer and the parts manufacturer are carried out independently of one other, then the implicit knowledge is often not converted to explicit knowledge due to lack of proper communication between the two. It is necessary to create and conceive new ideas which lead to new technologies by having inputs from the all the parties (component manufacturers, vehicle manufacturer and end users). This approach was used by Toyota to improve the reliability of the oil seal for the tansaxle.

To ensure high reliability of product design and quality assurance, a "Total Task Management Team" involving Toyota and NOK personnel was created to transform the implicit knowledge (relating to product and processes in both organizations) into explicit knowledge and to create new technologies of interest to both organizations. The "Dual Total Task Management Team" named "DOS-Q"(Drive-train Oil Seal-Quality Assurance Team: T Dos-Q5 and N Dos-Q8) formed between TMC and NOK is shown in Fig. 2.

Toyota's constituting teams comprise Q1 and Q2 in charge of investigation into the cause of the"oil leakage" and Q3 - Q5, which handled manufacturing problems relating to drive shafts, vehicles and transaxles. Similarly, NOK formed teams Q1 through Q8 as shown in Fig. 2. Q1 and Q2 at Toyota interacted closely with their counterparts at NOK to improve the reliability of the oil seals as a single unit and, likewise, Q3 - Q8 handled the manufacturing problems for quality assurance.

Accordingly, the teams shared their individual knowledge (relating to empirical techniques and other technical information) to apply them to solving the problems under consideration. Each team had a general manager and the joint team was led by Toyota's TQM Promotion General Manager for the vehicle. The methodology of "TDS-D" (Total Design System for Drive-train Development) involving IM (Information Management), TM (Technology Management) and PM (Production Management) was used as indicated in Fig. 2.

Study of the "Transaxle oil seal Leakage" mechanism using systematic approach "TDS-D"

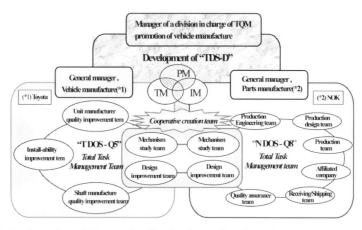

Fig. 2. Configuration of Cooperative Creation Team "Dual Total Task Management Team"

3.2. Problem Formulation and Task Setting

According to NOK, oil leaks occurred due to wear. The result of a wear test on oil seals indicates a running distance of 400,000 km (equivalent to 10 years or more vehicle life) is regarded as a sufficiently reliable design. The oil seal leakage from the failure repair history of the Toyota DAS (Dynamic Assurance System (see Sasaki, 1972) can be classified into initial failures that occur during the initial life of new vehicle and failures caused by wear that occur after running for some time. From the investigation by Q1 and Q2 teams, there are cases when oil leaks occur before reaching half of the running distance determined during an earlier test on oil seals alone. Thus, it cannot be said that the design is highly reliable and the failure mechanism is fully understood.

Judging from the survey and analysis of parts returned from customers for claims, the cause of the failure was identified as being due to the accumulation of foreign matter between the oil seal lip and the contact point of the transaxle shaft, resulting in insufficient sealing. Oil leaks were found not only during running, but also at rest. Thus, it was considered that the cause is poor foreign matter control during the process aiming at an improvement in production quality. It is considered that metal foreign matter in the oil in the transaxle gear box adversely affects the respective contact points, thus accelerating the wear of the oil seal lip during rotation of the axle shaft.

However, the permissible limit of the foreign matter particle size that causes oil leaks is unknown for the earlier failure, and the dynamic behavior resulting in oil leaks is not clarified yet for the latter failure. Consequently, the root of the problem is that the oil leak mechanism is not clarified and no quantitative analysis on cause/result correlation has been conducted, which is a true barrier for achieving high design reliability.

Task setting with all the people concerned, discussing the issues of the problems (and not depending on the rule of thumb practices), was done using affinity and/or association diagrams. These diagrams revealed that the essential problem was the lack of complete knowledge about the oil leak mechanism. It also reconfirmed that it was important to scientifically visualize the sealing phenomenon at the fitting contact portion. Consequently, it was decided that the engineering problems need to be solved by (1) clarifying the failure analysis processes through actual investigation of the parts, (2) carrying out a factor analysis of the oil leakage process, and (3) examine the design, manufacturing, and logistics processes involved.

4. Reliability Improvement of Oil Seal

4.1. The New Approach utilizing Science SQC

To clarify the oil seal leakage mechanism for transaxles, "Science SQC" (see Amasaka, 2000) was implemented with a mountain climbing type problem solution technique and using the "SQC technical methods" (see Amasaka, 1998), as indicated in Fig. 3. The three elements of "Management SQC" (see Amasaka, 1997), that is "Technology Management [TM]", "Production Management [PM]" and "Information Management [IM]" were developed as indicated in each stage of the team activity process.

As the figure illustrates, both teams are linked in the implementation of the"Management SQC" by combining the three management methods. In addition, the mountain climbing method for problem solution utilizing the"SQC Technical Methods" was used to achieve reliability improvement in terms of both design reliability and production quality. During the first year, the following technical themes were identified and summarized for future study by the teams:

(1) Why oil leaked from the oil seal?
(2) Had anybody actually seen the phenomenon?
(3) What was the oil seal design concept based on, while the axis is rotating or standing still?
(4) Was the oil leak attributable to the oil seal part or unit?
(5) Was the quality assurance system complete?

This led to the following actions.

(1) The market claim parts recovery method was improved and proper analysis was carried out for better understanding of the oil leak failure using "Information Management [IM]" in the second year of the study.
(2) A survey of production processes for the oil seal and transaxle unit using "Production Management [PM]" was done and a correlation analysis between market claims and in process rejects using picture mapping and the "Total QA Network" (see Amasaka and Osaki, 1999) was carried out for improving process control. Based on the knowledge acquired from these, a cause and effect analysis on the oil leak problem was conducted; fully utilizing a Science SQC approach, as well as "Technology Management [TM]", during the latter half of the second year to the first half of the third year. Concurrently, the oil

leak dynamic behavior visualization equipment was developed for analyzing and verifying the oil leak mechanism.

(3) The information management system was improved for systematic material acceptance inspection, production equipment maintenance, product inspection, shipment and logistics.

(4) In the latter half of the third year, design improvement and new technology design (quality improvement) were implemented in order to verify the effectiveness of the improvement and for horizontal deployment.

The remainder of the section gives more details of these activities that led to improving the design reliability and production quality.

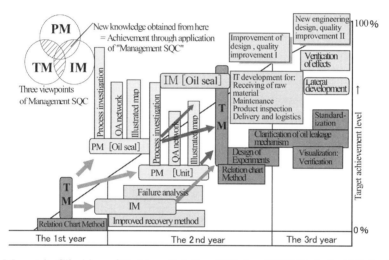

Fig. 3. Mountain Climbing of Problem Solution Utilizing "SQC Technical Methods"

4.2. Understanding of the Mechanism through Visualization

The study on the oil leak mechanism involved looking at different issues affecting the leak through several studies, as indicated in Fig. 4, involving the TM, PM and IT teams activities as shown in Fig. 3, in first and second year of the investigation. Fig. 4 covers various factors which could be acquired by investigation of TM activities based on the knowledge on various factors of PM and IT. To explore the causal relationship of oil leakage, we connected related factors with a (↓) mark using the relation chart method, and put the knowledge

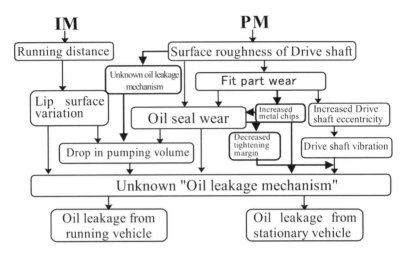

Fig. 4. Estimation of the Oil Leakage Mechanism

obtained so far into good order. But, because the oil leakage mechanism is unknown, the occurrence route of oil leakage on running and stationary vehicles isn't clear.

As indicated in the figure, for example, although the cause and effect relationships in the drive shaft surface roughness were recognized, it was not seen as being the main cause of premature leakage as opposed to wear, since the production process capability was ensured. However, this was not certain at that time. With new knowledge, it was deduced that the wear of the transaxle engagement part (differential case) during running increased the fine metal particles in the lubrication oil, which accelerated the wear of the oil seal lip unexpectedly. This in turn caused oil leakage due to a decrease in the sealing margin of the oil seal lip causing the oil seal pump quantity to drop.

The established theory used to be that fine metal particles (of micron order) would not adversely affect the lip sealing effect. When these are combined to produce relatively large particles, however, do they then affect the sealing effect? And what about the effect of alignment between the drive shaft and the oil seal (fixing eccentricity) during assembly? From another aspect, if oil leakage occurs due to foreign matter accumulation to the oil seal lip during transaxle assembly, what is the minimum particle size that causes the problem? These were unknown since the dynamic behavior of oil leakage was not yet visualized, so the true cause had yet to be clarified.

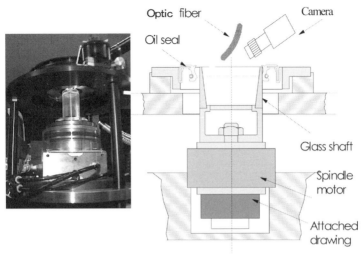

Fig. 5. Outline of device by visualization

Accordingly, a device was developed to visualize the dynamic behavior of the oil seal lip to turn this "unknown mechanism" into explicit knowledge, as shown in Fig. 5. As shown in the figure, the oil seal was immersed in the lubrication oil in the same manner as the transaxle, and the drive shaft was changed to a glass shaft that rotated eccentrically via a spindle motor so as to reproduce the operation in an actual vehicle. The sealing effect of the oil seal lip was visualized using an optical fiber.

It was conjectured that in an eccentric seal with one-sided wear, foreign matter becomes entangled at the place where the contact width changes from small to large. Two trial tests were carried out to ascertain if this was true or not. Based on the observation of the returned parts from the market and the results of the visualization experiment, it was observed that very fine foreign matter (which was previously thought as not impacting the oil leakage) grew at the contact section, as shown in Fig. 6 (Test - 1).

From a result of the component analysis, it was confirmed that the fine foreign matter was the powder produced during gear engagement inside the transaxle gear box. This fine foreign matter on top of microscopic irregularities on the lip liding surface resulted in microscopic pressure distribution which eventually led to the degrading the sealing performance (Test-2).

Also, the presence of this mechanism was confirmed from a separate observation that foreign matter had cut into the lip sliding surface causing aeration

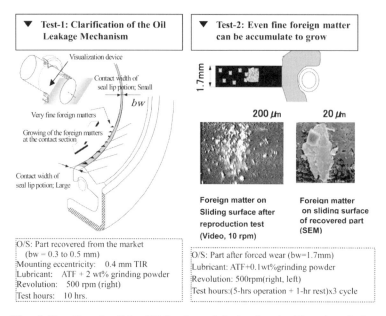

Fig. 6. Test Result of the Oil Leakage Mechanism by Test-1 and -2

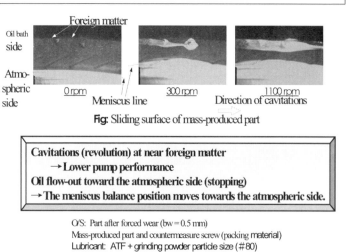

Fig. 7. Test Result of the Oil Leakage Mechanism by Test-3

(cavitations) to be generated in the oil flow on the lip sliding surface, thus deteriorating the sealing performance, as shown in Fig. 7 (Test-3). The figure indicates that cavitations occur in the vicinity of the foreign matter as the speed of the spindle increases, even when the foreign matter accumulated on the oil seal lip is relatively small.

This confirmed that such a situation leads to a reduction in the oil pump capacity of the oil seal, thus causing oil leakage during running. As the size of the foreign matter gets bigger, the oil sealing balance position of the oil seal lip moves more toward the atmospheric side and causes oil leaks at low speeds or when at rest. This was unknown prior to the study and hence not incorporated in the design of oil seals.

4.3. Fault and Factorial Analyses

Before studying the mechanism of oil leaks from oil seals described in Section 4.2, both NOK and Toyota considered that the wear of leaking oil seal lips would follow a typical pattern. The empirical knowledge based on the results of individual oil seal reliability tests was that the unit axle is highly reliable, and would ensure 400,000 km or more in B10 life (less than 10% of the items fail by B10). Because of smooth contact between the oil seal lip and the rotating drive shaft, due to a surface roughness with an oil film in between, it was thought that the oil seal lip should wear gradually.

As a result of the experiment discussed in Section 4.2, however, it was found that metal particles generated from gears in the differential case accelerates eccentric wear of the oil seal lip, making the expected design life unobtainable. Since the wear pattern was not simple, it had to be confirmed that faulty oil seals returned under claims would reproduce the oil leak problem.

For this purpose, a survey was conducted along with an experiment as indicated below:

First, in addition to defective oil seals, non-defective ones were collected on a regular basis to check for oil leak reproducibility and for comparison through visual observations. Next, transaxle units from vehicles, with and without oil leak problems, were collected on a regular basis to check for leak reproducibility in the same way. Integrating these results of transaxles with and without defective oil seals confirmed the defect reproducibility.

In all of these tests, oil leaks were reproduced as expected. Based on these test results, a Weibull analysis was conducted as described below.

The plot of the results (based on defective items resulting from claims) is shown in Fig. 8. It clearly shows a bathtub failure rate for oil seal failures. The three shape parameter (m) values correspond to three different failure modes. The figure resulted in the following new knowledge:

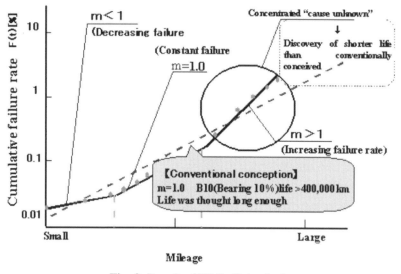

Fig. 8. Result of Weibull Analysis

(1) In the initial period, the failure rate is decreasing (slope (m) < 1), in the middle period it is constant (slope (m) = 1) and in the latter period it is increasing (slope (m) > 1) indicating a bathtub failure rate. The failure rate in each of the three section can be modeled by a different Weibull distribution, so that the failures can be modelled by a sectional Weibull model (See Blischke and Murthy, 2000).

(2) The initial failures (with failure rate decreasing) occur up to a run distance of 50,000km. Failures in the intermediate range (with failure rate constant) occur until 120,000km. Finally, failures occurring above this value (with failure rate increasing) are due to wear.

(3) The B10 mode life is approximately 220,000 km, about half the value stated as the design requirement.

To confirm the reliability of these results, the subsequent claims were analysed using the Toyota DAS system. Within the warranty period (number of years covered by warranty), the total number of claims classified by each month of production (total number of claims from the month of sales to the current month experienced on the vehicles manufactured in the same month) divided by the number of vehicles manufactured in the respective month of production is about twice the design requirement. This agrees with the result of the above reliability analysis.

The influence of five dominating wear-causing factors (period of use, mileage, margin of tightening, hardness of rubber and lip average wear width) was studied by two-group linear discriminant analysis using oil-leaking and non-oil-leaking parts collected in the past. The result showed high positive discriminant ratios of 92.0% and 91.7% for both group 1 (oil-leaking parts) and group 2 (non-oil-leaking parts). From the partial regression coefficients of the explanatory variables in the linear discriminant function obtained, the most significant influence was found to be the hardness of the rubber at the oil seal lip. The influence ratios of the five factors were obtained by means of an orthogonal experimental design (L27), with three level values, which were thought technically reasonable in consideration of the non-linear effects, assigned to each of them. (See Steinberg (1996) for a discussion of relevant experimental designs.)

Fig. 9 shows the influence ratios of each factor contributing to the

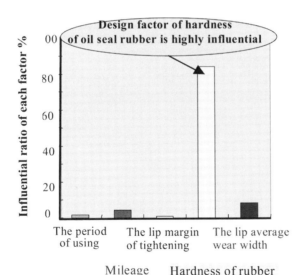

Fig. 9. Effect of Each Factor

discrimination. The figure shows that the hardness factor of the rubber is highly influential. This analytical result was also convincing in terms of inherent technologies. To test the validity of this result, the lip rubber hardness and the degree of wear of other collected oil seals was examined further. As a result, it has been confirmed that eccentric wear is more likely to shorten the seal life due because the rubber hardness at the lip portion decreases.

The result agrees with experience. Such survey and analysis could not have been carried out with the conventional separate investigation activities of Toyota or NOK. This could only be accomplished through the "Dual Total Task Management Team" activities between the two companies.

4.4. Design Changes for Improving Reliability

Based on the new knowledge acquired (as discussed in Sections 4.2 and 4.3), several design change plans were created to address the following two reliability problems: (1) The result of the Weibull analysis and visualization tests showed that some items had an unusually short life. It was recognized that it was to necessary to prolong the life of items. (2) The study confirmed that there was considerable variation in the oil seal lip hardness and this had to be controlled.

The following design modifications to improve the transaxle reliability were implemented:

(1) Improvement in wear resistance was achieved by increasing the gear surface hardness through changes to the gear material and heat treatment.
(2) The mean value of the oil seal lip rubber hardness was increased and the specification allowance range narrowed. This, in combination with improvements in oil seal lip production technology (including the rubber compound mixing process to suppress deviation between production lots), led to improving the process capability.

Some other issues not discussed here included the following. A design improvement plan was generated which involved a new lip form that prevented the reduction in the pumping capacity even when the sliding surface was worn out. The deposition of foreign matter (particularly when the lip was worn) was reduced by changing the design tolerance for the oil seal lip hardness and by improving the process capability. These together increased the B10 life to over

400,000 km. As a result, the cumulative number of claims by production month decreased by a factor of 10. The final outcome was achieving longer life, as initially planned.

4.5. Scientific Process Control

The goal of this activity was to further improve the reliability of oil seals and transaxles, produced in mass, by ensuring process capabilities and proper process management at both NOK and Toyota. This was achieved by controlling the quality variations needed to satisfy the specified design tolerances and a quality assurance management system to maintain the high quality. A new scientific process management, called "Inline-Online SQC" (see Amasaka and Sakai, 1996 and 1998) was used. This is capable of detecting abnormalities in the process that can affect product quality and preventing defective items being released through the use of "Science SQC" concepts.

A defect control monitor system was implemented for improving product reliability through intelligent process management and linking it to the IT (Information Technology) used in solving the oil leakage problem. Higher coaxial centers of metal oil seal housings, coil springs and seal lips, their alignment, contact width of oil seal lips, and thread profile identified during the design modification were properly monitored and controlled during the production to ensure high quality for the oil seals.

For transaxles, improvements in roundness and surface smoothness of the drive shaft (resulting in the reduction of metal particles that cause the wear of gears in the differential case) were achieved. Furthermore, the ARIM-BL (Availability and Reliability Information Manufacturing System-Body Line) was introduced, and this contributed to higher reliability in the process.

The quality assurance "QA Network" proposed and verified by the author (see Amasaka, 1998) was adopted and the "Toyota's Picture Maps system" was established to build quality and reliability into production processes from upstream to downstream and improve reliability through prevention of defective items being released. This was combined with the QC process chart, production chart and process FMEA (Failure Modes and Effects Analysis) that Toyota had been using for many years.

The quality failure picture map (called simply "Picture Map") was posted and used to clearly indicate the correlation between in-process quality failures and initial field failures (resulting in claims) in order to improve the quality awareness of operators at production sites and achieve complete conformance to the specified standards. For controlling foreign matter, the new information that the oil starts leaking with the presence of foreign matters of approximately 75 um in size (caused by the yarn dust produced from gloves during work, rubbish, powder dust, etc.) was publicized. This required the operators to prevent such material being introduced into the transaxle so as reduce early failures of the oil seal.

Similar "Picture Maps" were also used for material acceptance inspection and parts logistics as part of the total quality assurance system.

5. Conclusion

This case study deals with the new type of approach taken jointly by Toyota and NOK to improve the reliability of oil seal for transaxles of vehicles produced by Toyota. A "Dual Type Total Task Management Team" involving both Toyota and NOK was created to achieve the reliability improvement. This involved the use of "Management SQC", "SQC Technical Methods" and the core technologies of "Science SQC".

Toyota and NOK formed "TDOD-Q5" and "NDOS-Q8" teams to improve the reliability technology from Technology Management (TM), Production Management (PM) and Information Management (IM) viewpoints. Their R&D and design teams were closely linked to clarify the oil leakage mechanism and this led to the visualization technology for understanding the dynamics of the oil leakage. Based on this new knowledge and the Weibull analysis of failure data, the design of the seal was modified and higher reliability achieved. Through proper process management and "Total QA Network Activities", the claims from oil seal leak were reduced by a factor of 10.

At present, using the knowledge obtained from the studies done, development of the next generation oil seal is being carried out. It will have little wear and will prevent the leakage even if the lip is worn out.

Acknowledgement

We are indebted to Dr. Murthy, D. N. P., Professor of Department of Mechanical

Engineering, The University of Queensland, and Dr. Blischke, W. R., Professor of Department of Information and Operations Management, University of Southern California, for their support and comments on the original draft that resulted in a much improved article. We would also like to thank the persons concerned at NOK Corporation and Toyota Motor Corporation for their comments and suggestions.

References

Amasaka, K. (1997)."A Study on"Science SQC"by Utilizing"Management SQC"- A Demonstrative Study on A New SQC Concept and Procedure in the Manufacturing Industry-,"*Journal of Production Economics,* 60-61, 591-598.

Amasaka, K. (1998)."Application of Classification and Related Methods to SQC Renaissance in Toyota Motor,"*Data Science, Classification, and Related Methods, Springer,* 684-695.

Amasaka, K. (2000)."A Demonstrative Study of a New SQC Concept and Procedure in the Manufacturing Industry-Establishment of a New Technical Method for Conducting Scientific SQC-,"*The Journal of Mathematical & Computer Modeling,* 31, 1-10.

Amasaka, K., and Ootaki, M. (1999)."Development of New TQM by Partnering-Effectiveness of "TQM-SP" by Collaborating Total Task Management Team Activities-," (in Japanese) *Journal of the Society for Production Management, The 10 th Annual Technical Conference,* 69-74.

Amasaka, K., and Osaki, S. (1999)."The Promotion of New Statistical Quality Control Internal Education in Toyota Motor -A Proposal of"Science SQC" for Improving the Principle of TQM-," *The European Journal of Engineering Education on Maintenance, Reliability, Risk Analysis and Safety*, 24(3), 259-276.

Amasaka, K., and Sakai, H.(1996)."Improving the Reliability of Body Assembly Line Equipment," *International Journal of Reliability, Quality and Safety Engineering,* 3(1), 11-24.

Amasaka, K., and Sakai, H. (1998)."Availability and Reliability information Administration System "ARIM-BL" by Methodology in"Inline-Online SQC"," *International Journal of Reliability, Quality and Safety Engineering,*5(1), 55-63.

Blischke, W. R., and Murthy, D. N. P. (2000). Reliability: Modeling, Prediction, and Optimization, *Wiley, New York.*

Hirabayashi, H., Ohtaki, M., Tanoue, H. and Matsushima, A. (1979). "Troubles and Counter measures on Oil Seals for Automotive Applications," *SAE Technical paper series,* 790346.

Kameike, M., Ono, S. and Nakamura, K. (2000). "The helical seal: Sealing concept and rib design," *Sealing Technology, International, Elsevier* (77), 7-11.

Kawahara,Y., Abe, M., and Hirabayashi, H. (1980). "An analysis of sealing characteristics of oil seals," *ASLE Transactions,* 23(1), 93-102.

Lopez, A. M., Nakamura, K., and Seki, K. (1997). "A study on the sealing characteristics of lip seals with helical ribs,"*Proceedings of the15th International Conference of British Hydromechanics Research Group Ltd 1997 Fluid Sealing:* 1-11.

Nakamura, K. (1987). "Sealing mechanism of rotary shaft lip-type seals," *Tribology International.,* 20(2), 90-101.

Sasaki, S. (1972). "Collection and Analysis of Reliability Information on Atomotive Industries," *The 2nd Reliability and Maintainability Symposium, JUSE (Union of Japanese Scientists and Engineers):*385-405.

Sato, Y., Toda, A., Ono, S and Nakamura, K. (1999). "A study of the sealing Mechanism of radial lip seal with helical ribs -Measurement of the lubricant fluid behavior under sealing contact," *SAE Technical Paper Series,*1999-01-0878.

Steinberg, D. M. (1996). "Robust designs: Experiments for improving quality," *Chapter 7 in Ghosh, S., and Rao, C. R. (Eds.), Handbook of Statistics,* 13, *North-Holland, Amsterdam.*

Chapter 12: CAE Design

Improvement in CAE Analytic Precision:

A Study of Estimating Coefficients of Lift at Vehicle
- Using Neural Network and Multivariate Analysis Method Together-

It is shown that combining multivariate analysis with neural networks is useful for solving problems with complex interaction such as vehicle aerodynamics of lift. This method leads us to various valuable observations. In this paper, we show the effectiveness of this method.

Regarding some of past insufficiencies of solving problems by multivariate analysis method or neural networks only, there are conditions such as (1) difficulty in data collection, (2) mere know-how and (3) predicting complex interaction. By combining neural networks method with multivariate analysis, however, we can solve these problems efficiently. In addition, solving these problems by combining neural networks method with multivariate analysis, the works of hidden layers in neural networks can be found.

Keywords; Estimationg Coefficients of Lift, Vehicle, Neural Network, Multivariate Analysis Method.

1. Introduction

Increasing vehicle speeds in recent years have been calling for improvement of straight line stability during high-speed traveling. One of the means to improve the straight line stability is to reduce the lift. It is necessary to estimate the vehicle aerodynamics of lift in the design planning stage for realizing a profile having less lift. Special aerodynamic models prepared and used so far for experimentally estimating the lift in the design planning stage, however, have been insufficient for obtaining accurate data.

Furthermore, the modeling cost and experimental man-hours were both heavy burdens. In other words, we have been looking for developing a simple lift estimation method. While many examples of calculation [1-6] based on CAE analyses have been reported, the accuracy of estimation has yet to be improved for satisfactory vehicle development.

Recently, CAE analysis [7-8] with neural networks (hereinafter abbreviated as NN) has been used. NN applications have been limited to introducing the function that expresses the non-linear input-output relationship when factors explaining the causality can almost be identified [7], and to estimating the unknown causality when multiple analytical data are available [8].

This paper proposes combined use of the neural networks and multivariate analysis for analyzing the unknown causality concerning the vehicle aerodynamics of lift and reports a study to demonstrate its effectiveness in improving the lift estimation accuracy.

2. Aerodynamics of Lift

Fig. 1 illustrates a large wind tunnel used for measuring the lift of an actual vehicle. Load cells are embedded at the tire grounding points in the wind tunnel to enable the longitudinal, vertical and transverse forces to be measured. The lift is expressed as the vertical force applied to the load cell, and is generally measured at a wind velocity equivalent to 120 km/h. Generally, the lift is separated into the forces applied to the front and rear wheels (*Lf* and *Lr*) as shown in the figure, which are expressed as zero-dimension coefficients [9], *CLf* and *CLr* in equation (1).

$$\left. \begin{aligned} CLf &= \frac{Lf}{\frac{1}{2}\,\rho V^2 A} \\[2ex] CLr &= \frac{Lr}{\frac{1}{2}\,\rho V^2 A} \end{aligned} \right\} \quad (1)$$

CLf: Front wheer lift coefficient
Lf: Front wheel lift [N]
ρ : Air density [kg/m³]
A: front projection area [m²]

CLr: Rear wheel lift coefficient
Lr: Rear wheel lift [N]
V: Vehicle speed [m/s]

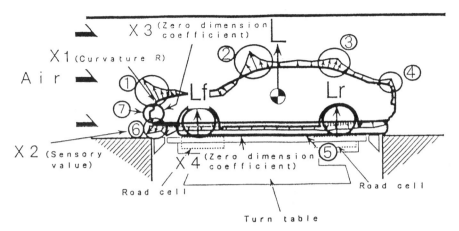

Fig. 1. Condition in a wind tunnel and pressure distribution of vehicle's center

Since the lift coefficient is estimated in the same way for front and rear wheels, that for front wheels is discussed herein.

3. Combined Use of NN and Multivariate Analysis

3.1. NN Characteristics

There are some engineering problems requiring analysis of non-linear correlation between complicatedly entangled factors. In solving such problems, a BP (back propagation) type NN having one middle layer is used recently for modeling the non-liner structure [10-11]

It is difficult to analytically obtain strict scientific rules by modeling. If objective and explanatory variables that physically explain the causality can be introduced, the NN can realize, through learning of cases, a generalized model can be formulated by patterning the regularity of input-output characteristics, which has so far been difficult.

With the BP method, the calculation time increases and convergence becomes extremely poor because of an increasing number of meaningless input variables resulting from lack of careful selection of essential explanatory variables. On the other hand, technically complicated and hard-to-understand non-linear model may be used for identification if the convergence condition is determined severely without successful introduction of technically necessary variables since learning is to be based only on introduced variables.

Consequently, technically meaningless, strained causality will be identified if learning is converged only on the basis of known variables introduced empirically. As a result, failure in identification of the originally desired unknown variables will occur. Especially, in immature engineering fields, robust calculation results cannot be obtained, resulting in lack of appropriateness and reliability.

3.2. Characteristics of Multivariate Analysis

Multivariate analysis is useful in extracting the potential structure and rules hidden in multivariate data. Among many methods, the multiple regression analysis is useful for formulating the causality [12-14] by selecting appropriate variables while technically and carefully interpreting the basic statistics as well as histograms and scatter diagrams. Formulation using a multiple regression model will bring about overall fitness as seen from the contribution rate.

The correlation between explanatory variables (with or without interaction), which is the problem in the multiple regression analysis, can be checked by means of the VIF value showing the multicollinearity as well as the sign and value of the partial regression coefficient. Combined use of residual analysis and partial residual regression plot will enable the non-linear effect of each explanatory variable and any necessary unknown explanatory variables to be sought as stratification factors. When the following analytical methods are used together, multiple regression analysis offers high flexibility.

When non-linear characteristics are observed, linear modeling with curve fitting and variable conversion is required. When stratification factors are observed, the influence from categorical qualitative variables should be checked and replacement with quantitative variables be evaluated. When interaction between explanatory variables is also observed, addition of a new explanatory variable may be needed.

Especially when complex causality exists, sequential manual operations iterative search and technical evaluation are required in order to attain technically appropriate and highly reliable factorial analysis and regression modeling. The problem with this method is the need of excessive time and labor.

3.3. Combined Use of NN and Multivariate Analysis

Clarifying existence of non-linear characteristics and complicated causality is a problem with lift characteristic estimation as an immature engineering field as described in section 1. When there is a limited number of available data, technically consistent causality can be determined by technically searching existence of any unknown variables hard to be extracted merely by experience and skill or by extracting any interaction between explanatory variables. As a methodology, we proposing combined use of NN and multivariate analysis which interpolate and assist each other.

First, conduct the factor analysis based on the empirically obtained variables and then search or confirm existence of unknown variables by making use of the features of multiple regression analysis, to determine necessary variables. For this purpose, the contribution rate, residual analysis and partial residual regression plot will be effective. Next, complicated causality is evaluated as a technically meaningful model, utilizing the learning ability of the NN so as to identify a generalized behavioral model.

When a practically convenient, generalized model based on physical phenomena is to be made this way for causal analysis aimed at improving the lift estimation accuracy, the approach should not be from empirical formula to NN analysis but from factorial analysis using an empirical model based on multivariate analysis to NN analysis. The effectiveness of this approach is proven below.

4. Analysis Concerning Improvement of Lift Estimation Accuracy

4.1. Evaluation by Empirical formula

4.1.1. Preparing Empirical formula

Lift estimation in the product development stage should be able to be conducted based on the vehicle profile data. The *CLf* is influenced by an envelope factor such as the front wheel lift and front projection area (forward projected area from the front of vehicle) as shown by equation (1). So, the conventional study [9] could not identify the correlation between the profile data and the lift of each vehicle, which was inconvenient for the design for *CLf* estimation and control.

Therefore, we decided to select relatively more influential factors and create a technically more convenient empirical formula to check the causality between the profile data and the lift.

First, the aerodynamic force applied to the vehicle is checked in order to generate an empirical formula. The ↑ mark shown in Fig. 1 represents the distribution of the pressure to the center of vehicle. It can be seen from the figure that the pressure is applied to ① leading edge of hood, ② front edge of roof, ③ rear edge of roof, ④ rear edge of luggage compartment, ⑤ entire floor bottom and ⑥ bottom of front bumper. Then, the air flow loss by engine cooling wind ⑦ is added, which is empirically known as a very influential factor. On the other hand, factors ②, ③ and ④ are removed as they are less influential, also from experience. Thus, the following four factors are selected:

$X1$: (Curvature R, ① in the figure)
$X2$: (Sensory value on separation* strength, ⑥ in the figure)
$X3$: (Dimensionless resistance coefficient, ⑦ in the figure)
$X4$: (Dimensionless resistance coefficient, ⑤ in the figure)
(* Represents separation of air flow.)

Data shown in Fig. 2 are collected from 38 vehicles ranging from popular 4-door sedans to medium- and high-class vehicles to generate an empirical formula.

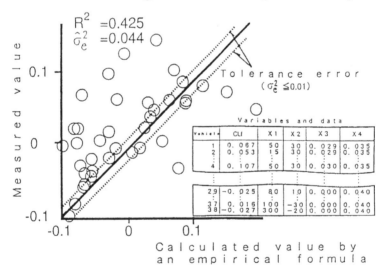

Fig.2. A correspondence between measured value and calculated value by an empirical formula

With regard to $X1$, the smaller the curvature (R), the greater the resulting lift. Since the increase is greater in the region where curvature R is small, an inverse function is assumed.

The coefficient is determined by estimating the ratio against the entire lift coefficient based on the experience. Similarly, the value of $X3$ is determined by estimating the ratio against the entire lift coefficient empirically obtained. With regard to $X2$, the separation strength rank is determined by sensory evaluation by visualizing the air flow. It is assumed that the influence of $X4$ affects factors $X1$, $X2$ and $X3$. CLf equation (2) shown below is generated.

$$CLF = (3.6 \times X3 - 0.01 \times 2 + \frac{2.5}{X1} - 0.075) \times \frac{X4}{0.03} \quad (2)$$

The reason why the product of multiple of $X4$ is considered in equation (2) is because $X4$ is empirically judged to influence the overall vehicle. As the practical value of $X4$ for actual vehicles is 0.02 to 0.04, $X4/0.03$ is practically close to 1. As a result of a technical study based on the pressure distribution shown in Fig. 1, it is a general method for engineers to extract $X4$ from $X1$ as an alternate value of CLf and to replace the causality with CLf with an empirical value by modeling.

4.1.2. Calculation Result

The scatter diagram in Fig. 2 shows the correlation between the actually measured lift values and the values calculated from the measured data using equation (2). The contribution rate (hereinafter referred to as R^2) is calculated as $R^2 = 0.43$. As seen from the figure, the values of many vehicles are out of acceptable tolerance. Although the reason for the departure from the tolerance cannot be explained from the engineering standpoint, it is considered that the coefficient in equation (2) is inappropriate.

However, we considered from the result that it would be possible to estimate the lift by selecting all necessary factors without omission and by modeling the causality through deeper technical analysis.

4.2. Evaluation by NN Analysis

4.2.1. Calculation Result

We have tried NN analysis on assumption that sufficient accuracy from the engineering viewpoint would be obtained even by use of only four profile factors (herein called the explanatory variables) $X1$ through $X4$, if the coefficient of empirical formula is optimized and the interaction between explanatory variables is taken into consideration.(Fig. 3)

Since the number of sampled data is small and number of explanatory variables is limited at 4, the three-layered structure [15] with one middle layer is adopted as shown in Fig. 3 (a). Three middle units are used based on our experience concerning convergence [7-8]. The degree of correlation between units is expressed by the thickness of each solid line. The calculation is conducted using the back propagation method and at 200 learning cycles. Of the 38 test vehicles, 28 is used for the study and remaining 10, for verification.

Fig. 3 (b) shows the result of NN analysis. Contribution rate R^2 was 0.96 while estimated error σe^2 was 0.012, showing satisfaction of the required accuracy ($\sigma e^2 = 0.01$ or less).

4.2.2. Result Verification and Consideration

Technical consistency of the result of NN analysis has been tested. When the influence of $X3$ obtained by calculation is compared with a separately experimental result on the individual effect of $X3$, for instance, the results of the experiment and NN analysis are different as shown in Fig. 3 (c).

The reason for technical contradiction between the results of experiment and NN analysis may have been caused by excessive learning due to the small number of tested vehicles ($n = 28$) for four explanatory variables ($X1$ to $X4$) with probable omission of other higher explanatory factors. This has brought about a high contribution rate ($R^2 = 0.96$), leading to a difference from the experimental result. The same applies to other explanatory variables.

Either or both of the following requirements should be satisfied for improving the reliability of NN analysis to obtain technically consistent results:

(1) Improving explanatory variables by technically selecting those with high

225

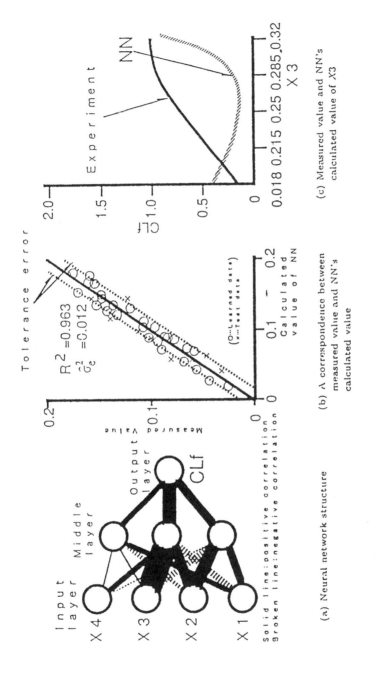

(a) Neural network structure

(b) A correspondence between measured value and NN's calculated value

(c) Measured value and NN's calculated value of X3

Fig. 3. Analysis of neural network

contribution rates

(2) Increasing the number of samples so as to obtain data at least 2 to 3 times the number of explanatory variables or, if possible, 5 times or more judging from our experience

(3) It is necessary to combine both either or both

All technically significant explanatory variables should be selected without omission in order to improve the explanatory variables. In modeling for lift coefficient estimation with high accuracy in a field involving limited experience as described in Section 2, it is necessary to find out unknown explanatory variables and to carefully select those to be used. We have, therefore, considered that multivariate analysis tested and reported sufficiently so far [16-21] would be effective for solving engineering problems mainly through factorial analysis.

4.3. Evaluation by Multivariate Analysis

4.3.1. Analysis of Data Used in Empirical Formula

Table 1 shows the effectiveness of explanatory variables $X1$ to $X4$ obtained by multiple regression analysis by designating all variables using the quantitative data used for calculation of empirical formula in Fig. 2. As in the case of Fig. 2, the obtained contribution rate was not satisfactory at $R^2 = 0.71$.

When looking at the standard partial regression coefficient that represents influence of each explanatory variable, the influence of $X1$ was small. In addition, the sign of $X4$ coefficient is negative. Thus, there are many points that do not agree with empirical knowledge. Furthermore, there may be other explanatory variables omitted.

4.3.2. Finding out Factors

What is important in selecting factors and improving the analytical accuracy is how to find out empirically unknown factors. Histograms and scatter diagrams were effective as one analytical process. A bi-variate scatter diagram, with possible existence of stratification factor, extracted from histograms and scatter

Table 1. A study of 4 factor's effect by an empirical formula

A study of four factors (x1~x4) (linear multivariate analysis)

Multiple correlation	R	0.8403		Adjusted R-square	R*2	0.6704
R-square	R²	0.7061				

**************Analysis of variance table**************

	Degrees of freedom	Sum of square	Mean square	F-value
Regression	4	0.0866	0.0217	19.8183**
Residual	33	0.0361	0.0011	
Total	37	0.1227		

Factor's

No	Stand.BETA	Error of BETA	F	VIF
X1	−0.1192	0.1309	0.8285	1.9240
X2	0.2120	0.1353	2.4553	2.0543
X3	0.7533	0.0945	63.5659	1.0023
X4	−0.2103	0.1302	2.6075	1.9034

X2   X4

X2   X3

(Plots are overlapped)

diagrams is shown in Table 1. For example from $X2$ and $X4$ scatter diagrams, the stratification factor of $X2$ is considered to be the portion influenced by the flow at the bottom of the bumper layer receives the influence from the flow at the bottom of the bumper.

As a result of our technical evaluation, the effect of the tire profile ($X5$) has been found as an influential factor. Similarly from $X2$ and $X3$ scatter diagrams stratified into three groups, the overall vehicle length ($X6$) related to the vehicle size has been found as an influential factor in view of the physical phenomenon of airflow. Is seems that these can be converted into quantitative variables from the engineering viewpoint.

To improve the analytical accuracy, profile factors of the front bumper bottom that governs the sensory value of $X2$ were technically checked from the aerodynamic viewpoint. Seven seemingly influential factors ($X21$ through $X27$), including the sectional curvature (R) at the front bumper bottom, would be significant. These seven factors should include $X4$ from the engineering standpoint.

Similarly, an engineering analogy based on the factorial effect of $X1$ which represents the curvature at the front edge of the engine hood indicated that the distance from the crest of the curvature (R) at the front edge of the engine hood to a specific point may be replaced with $X1$ as its quantitative factor. Factors $X5$, $X6$ and $X7$ have been confirmed influential by other experiments. As for the causality between $X2$ and $X21$-$X27$, the data in Table 1 were used again to obtain quantitative data for seven profile factors $X21$ through $X27$.

The contribution rate obtained by multiple regression analysis with designation of all variables was $R^2 = 0.91$, indicating that the sensory value of $X2$ can be replaced with quantitative variables. Judging from the values of standard partial regression coefficients, $X21$ and $X27$ are less influential while remaining profile factors $X22$ through $X26$ are influential with technical significance.

4.3.3. Execution of Multiple Regression Analysis

Table 2 shows the data sheet (a) for multiple regression analysis using selected factors and the result (b) obtained by increasing and decreasing the variables. The contribution rate was $R^2 = 0.92$, indicating better explanation in lift coefficient (CLf) estimation. Based on the standard partial regression coefficients, influences of $X3$ and $X5$ were confirmed. Then, $X22$, $X23$, $X25$ and $X26$ as substitution of $X2$ as well as $X7$ and $X8$ were selected.

We consider, however, that unselected $X24$ is not negligible from the technical viewpoint considering interaction between profile factors. Fig. 4 shows the example of partial residual regression plot executed to obtain the degrees of influence of individual variables on CLf. Each explanatory variable was able to be evaluated more deeply. One example is that $X3$ and $X5$ show non-linear instead of linear effects as shown in the figure, which agrees well with technical estimation.

As described above, new empirically unknown profile factors have been selected thanks to the multivariate analysis. It is considered that almost all necessary explanatory factors are considered to be selected according to their contribution rates obtained by multiple regression analysis only. Furthermore, the degree of influence of each explanatory variable for causality explanation has been identified technically. Thus, linkage with NN analysis enabling lift estimation modeling has been attained, which is effective for searching interaction between

Table 2. A data and end result

(a) Variables and data

		Dependent variable	Independent variable											
		CLf	X3	X5	X6	X7	X21	X22	X23	X24	X25	X26	X27
Compact	(1)	-0.028	0.000	12.450	44.000	710.000	1.000	75.000	200.000	248.000	10.000	360.000	80.000
	(2)	-0.040	0.000	12.450	44.000	710.000	0.000	55.000	200.000	250.000	10.000	360.000	80.000
	(3)	-0.043	0.000	12.450	50.000	690.000	2.000	50.000	200.000	262.000	10.000	345.000	100.000
	(4)	-0.017	0.000	13.964	50.000	690.000	2.000	50.000	200.000	262.000	10.000	345.000	100.000
	(5)	-0.005	0.000	13.964	50.000	677.000	2.000	60.000	200.000	245.000	0.000	345.000	100.000
	(6)	-0.101	0.000	12.450	43.000	685.000	0.000	40.000	250.000	250.000	50.000	360.000	70.000
	(7)	-0.087	0.000	12.450	40.000	700.000	0.000	40.000	250.000	260.000	50.000	360.000	70.000
Medium	(8)	:	:	:	:	:	:	:	:	:	:	:	:
		:	:	:	:	:	:	:	:	:	:	:	:
	(30)	0.087	20.494	12.845	52.000	710.000	0.000	70.000	200.000	265.000	40.000	360.000	125.000
	(31)	0.066	14.832	12.845	47.000	710.000	0.000	55.000	200.000	265.000	40.000	360.000	125.000
Large	(32)	0.075	14.832	12.845	52.000	740.000	0.000	85.000	200.000	259.000	0.000	290.000	30.000
	(33)	0.087	15.166	13.964	50.000	750.000	0.000	50.000	200.000	244.000	70.000	490.000	150.000
	(34)	0.062	14.142	13.229	42.000	690.000	0.000	40.000	200.000	260.000	0.000	360.000	65.000
	(35)	0.070	16.125	13.602	60.000	720.000	1.000	20.000	100.000	295.000	40.000	400.000	125.000
	(36)	0.093	16.733	13.964	60.000	720.000	1.000	20.000	100.000	267.000	35.000	370.000	120.000
	(37)	0.085	15.492	13.964	60.000	715.000	0.000	15.000	100.000	295.000	0.000	290.000	35.000
	(38)	0.029	13.416	14.318	70.000	680.000	2.000	25.000	3.333	250.000	25.000	550.000	90.000

(b) Reasult of multivariate regression

Multiple correlation R-square	R 0.9600 R² 0.9216		Adjusted R-square R*² 0.9000	
★★★★★★★★★★★Anâlysis of variancê tablè★★★★★★★★★★★★★				
	Degrees of freedom	Sum of square	Mean square	F-value
Regression	8	0.1130	0.0141	42.6341★★
Residual	29	0.0096	0.0003	
Total	37	0.114		

Factor's No.	Stand.BETA'	Error of BETA'	F	VIF
X3	0.7888	0.0520	229.8700	1.0016
X5	0.4698	0.0720	42.6339	1.9162
X6	-0.1189	0.0709	2.8159	1.8588
X7	0.1600	0.0644	6.1740	1.5351
X22	0.1926	0.0781	6.0829	2.2576
X23	0.1993	0.0742	7.2113	2.0387
X25	-0.1409	0.0596	5.5974	1.3128
X26	-0.1542	0.0610	6.3922	1.3773

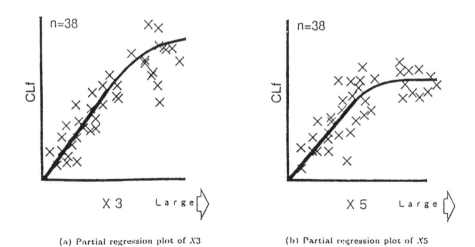

(a) Partial regression plot of X3   (b) Partial regression plot of X5

Fig. 4. Re-analysis of influential factor

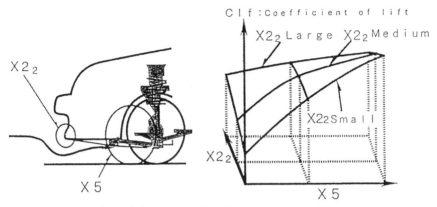

Fig. 5. The interaction between $X5$ and $X22$

arbitrary non-linear functions and profile factors.

One difficult point with this analytical method is to determine the correlation between profile factors. Because of aerodynamics, it is predicted that the down-stream profile factors are affected by up-stream profile factors. For example, our experience suggests interaction between $X22$ and $X5$ as shown in Fig. 5. It is almost impossible to find out all combinations of interactions between profile factors without omission based on the present empirical engineering and the characteristics of multivariate analysis. NN analysis capable of analyzing correlation between all factors is required in the next analysis stage.

4.4. Re-analysis by NN

4.4.1. Calculation result

Fig. 6 (a) illustrates the circuitry calculated after re-learning based on all eleven explanatory variables and the knowledge obtained by multivariate analysis as well as the data sheet in Table 2 (a). The circuitry is 3-layered with one middle layer. Evaluations are made sequentially by sequentially changing the number of neurons (h) in the middle layer between 2 and 7 ($h = 2$ to 7). The contribution rate (R^2) in terms of correspondence between the measured and calculated values is high, the estimation error (σe^2) is small and the result is technically reasonable when the number of neurons is 3. The number of variables selected similarly as in the multivariate analysis in 4.3 is nine and all of them are identical.

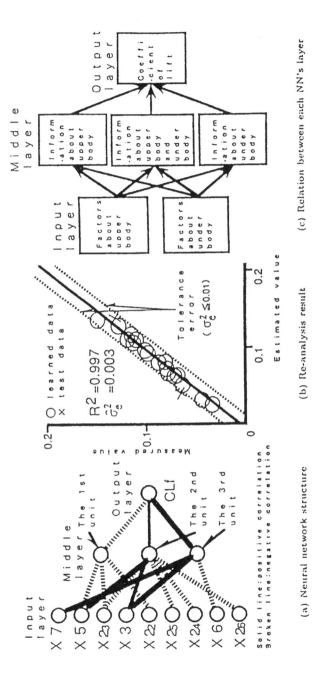

(a) Neural network structure

(b) Re-analysis result

(c) Relation between each NN's layer

Fig. 6. Re-analysis of neural network

The thickness of each line in the figure represents the incidence number (only when it is large). Solid lines indicate positive correlation while dotted lines indicate negative correlation. As a result of NN analysis, $X3$ and $X5$ have been found to have especially high contribution rates, showing agreement with the result of multivariate analysis. From the calculation result shown in Fig. 6 (b), R^2 = 0.99 to indicate good matching. $\times$ marks indicating estimation without learning show practically sufficient accuracy, suggesting applicability to new product development in the future.

4.4.2. Consideration

The NN analysis has taken complicated interactions into consideration as described in 3.1. Concerning the functioning of the middle layer in the NN, it is still under study by many researchers [10-11]. When looking at the learning circuitry in Fig. 6 (a) after the calculation obtained by this study, it is understood that the three units in the middle layer in Fig. 6 (c) are related to the vehicle upper profile, floor bottom profile, and both the upper and floor bottom profiles.

It is supposed that the middle layer works as a relaying terminal which bundles common characteristics of individual factors into larger categories through information transfer between correlated factors. With regard to the correlation between $X22$ and $X5$, the estimation result in Fig. 6 has been verified through NN analysis. When $X22$ is large, separation of the air flow under the front bumper increases to lessen the influence from the airflow received by $X5$. This agrees satisfactorily with the actuality.

5. Conclusion

(1) The effectiveness of combining the NN and multivariable analysis has been proposed for studying the causality of vehicle aerodynamics of lift.

(2) Practical examples where identifying the causality is difficult by either conventional NN analysis or multivariate analysis because of the difficulty in sampling the data, empirical technologies are insufficient, causality is non-linear or interactions between explanatory variables exists have been shown through analytical study on causality to verify combined use of the two analyses is effective for efficient solution of engineering problems.

(3) The following conclusions are obtained through practical study for verifying the effectiveness of combined use of NN and multivariate analysis:

① If the number of samples is small, NN analysis based merely on empirical technology may bring about high correlation even when the degree of influence and factor selection are insufficient. in such a case, evaluation from technical viewpoint should be made.

② When empirical technology is lacking or the available number of data is small, multivariate analysis will permit a study based on already sampled data or reveal new factors with less test experiments.

③ An optional non-linear function or interaction can be taken into consideration by NN analysis of the results in ② above to obtain new knowledge hard to get empirically or experimentally.

(4) The engineering analysis process by making the most of the NN and multivariate analysis which interpolate each other plays an important role in analyzing and evaluating the aerodynamics of lift. A further study on testing the methods and results so far obtained will contribute to shortening the research and development period.

References

[1] Yamada, Itoh, et al: Calculation Analysis on Vehicle Aerodynamics; (in Japanese) *Journal of Society of Automotive Engineers of Japan, Report at Academic Lecture Meeting*, 921, No. 921004 (1992).

[2] Tanaka, Himeno, et al: Application of Numeric Aerodynamics Analysis System for Aerodynamics Development; (in Japanese) *Journal of Society of Automotive Engineers of Japan, Report at Academic Lecture Meeting,* 921, No. 921002 (1992).

[3] Kataoka, Ukita, et al: Numeric Analysis on Vehicle Aerodynamics of Lift Including Floor Bottom and Engine Compartment; (in Japanese) *Journal of Society of Automotive Engineers of Japan, Report at Academic Lecture Meeting,* 921, No. 921003 (1992).

[4] Satoh, Shibazaki, et al: Vehicle Aerodynamics Analysis Using Surface Automatic Grating Producing Method; (in Japanese) *Journal of Society of Automotive Engineers of Japan, Report at Academic Lecture Meeting,* 931, No. 9301601 (1993).

[5] Hashiguchi, Kuwahara: Numeric Analysis for Vehicle Aerodynamics; (in Japanese) *Journal of Society of Automotive Engineers of Japan, Report at Academic Lecture Meeting,* 931, No. 930153 (1993).

[6] Kuriyama: R&D of Numeric Aerodynamics and Utilization in Designing; (in Japanese) *Journal of Society of Automotive Engineers of Japan, Report at Academic Lecture Meeting,* 921, No. 921001 (1992).

[7] Kageyama: Tire Load Ratio Control Using Neural Network; (in Japanese) *Journal of Society of Automotive Engineers of Japan, Report at Academic Lecture Meeting,* 924, No. 924159 (1992).

[8] Morita: Combustion Parameter Optimization Control for Gasoline Engine Using Neural Network; (in Japanese) *Journal of Society of Automotive Engineers of Japan, Report at Academic Lecture Meeting,* 924, No. 924001 (1992).

[9] W. Hucho: Aerodynamics of Road Vehicles, *Butterworths* (1987).

[10] Yakawa: Application to Neural Network Calculation and Applied Aerodynamics; (in Japanese) *Baifu-kan* (1992).

[11] K.Funabashi: Neural Network Technology; (in Japanese) *Plasticity and Processing,* Vol. 34, No. 387, pp. 352-357 (1993).

[12] T.Takaoka, K. Amasaka: Analysis on Fuel Economy Improving Factors; (in Japanese) *Quality, Journal of the Japanese Society for Quality Control,* Vol. 21, No. 1, pp. 64-69 (1991).

[13] K.Kusune, K.Amasaka, et al: Overall Curvature of Press Parts Having Large Curvature, Spring Back Analysis; (in Japanese) *Quality, Journal of the Japanese Society for Quality Control,* Vol. 22, No. 4, pp. 24-30 (1992).

[14] K.Amasaka, Y. Mitsuya, et al: Study on Rust-proof Quality Assurance for Plated Parts Using SQC; (in Japanese) *Quality, Journal of the Japanese Society for Quality Control,* Vol. 23, No. 2, pp. 90-98 (1993).

[15] K. Hunabashi: On the approximate realization of continuous mappings by neural networks: *Neural Networks,* Vol.2, No.3, pp.183-192 (1989).

[16] K. Amasaka: SQC Development and Effect at Toyota; (in Japanese) *Quality, Journal of the Japanese Society for Quality Control,* Vol. 23, No. 4, pp. 47-58 (1994).

[17] K.Amasaka: A study on "Science SQC" by Utilizing "Management SQC,- A Demonstrative Study on a New SQC Concept and Procedure in the Manufacturing Industry-: *Journal of Production Economics,* Vol.60-61, pp.591-598 (1999).

[18] K.Amasaka, and S.Osaki: The Promotion of New Statistical Quality Control Internal Education in Toyota Motor -A Proposal of "Science SQC" for Improving the Principle of TQM-:, *European Journal of Engineering Education (EJEE) , Research and Education in Reliability, Maintenance, Quality Control , Risk and Safety ,*Vol.24, No.3, pp.259-276 (1999).

[19] K.Amasaka: A Demonstrative Study of a New SQC Concept and Procedure in the Manufacturing Industry, -Establishment of a New Technical Method for Conducting Scientific SQC-: *An International Journal of Mathematical & Computer modeling,* Vol.31,No.10-12, pp.1-10 (2000).

[20] K.Amasaka and S.Osaki: A Reliability of Oil Seal for Transaxle - A Science SQC Approach in Toyota-: *Case Studies in Reliability and Maintenance by Wallace R.Blischke and D.N.P.Murthy, to be published by John Wiley & Sons, Inc.,* pp.571-581 (2002).

[21] K.Amasaka: Proposal and Implementation of the "Science SQC" Quality Control Principle, *International Journal of Computer Modeling,* Vol. 38, No. 11-13, pp.1125-1136, (2002).

Chapter 13: Production Engineering

Spring Back Control Measures:

Analysis of Total Curvature Spring Back in Stamped Parts with Large Curvature

This study is to examine the effectiveness of "Science SQC" approach methods in total curvature spring back for gaining quantitative data that supports engineering. In the first step, we conducted model experiments in a test plant and identified the factors that may affect spring back. We obtained data using the production equipment for prototype, conducted multiple regression analysis and established a formula to estimate the amount of spring back. In the second step, we followed a similar analysis process to develop the control measures for spring back in the mass production stage and verify our analysis with the production equipment to estimate the effects of the measures.

Keywords; Total Curvature Spring Back, Stamped Parts, Large Curvature, "Science SQC" approach, Vehicle body parts.

1. Introduction

A vehicle body consists of many stamped parts that are formed from thin steel plates. In order to attain high quality of the body in areas such as rigidity and buildability, high dimensional precision is required for each component. Generally speaking, precision problems are likely to occur in thin wall stamped parts because the parts are subject to elastic recovery, or spring back (or SB), after the forming process. The present engineering technology is yet to be capable of overcoming this problem.

In terms of parts with unified cross-sectional shape, there have been a number of theoretical and empirical analyses for the relation between spring back and the different stresses in the cross-sheet direction, and the methods to control spring back have been almost established. When the parts have a complex shape, as in the case of vehicle body parts, spring back and torsion occur in all portions of the parts.

Spring back in this case is generated by a combination of factors: not only the cross sectional stress but also the stress on the sheet surface and shape. Due to such complex factors, there have been few technical studies to identify the causes and establish control measures as shown in Table 1. With the current development of computer technology, CAE, especially FEM analysis, is advancing remarkably. Although CAE has entered the maturity stage, it still cannot be utilized for controlling spring back. With this background, estimation of potential problems and incorporation of the solution at the die engineering stage still depend on the know-how of skilled engineers as shown in Table 2.

The purpose of this study is to examine the effectiveness of "Science SQC" approach methods in total curvature spring back for gaining quantitative data that supports engineering. In the first step, we conducted model experiments in a test plant and identified the factors that may affect spring back. We obtained data using the production equipment for prototype, conducted multiple regression analysis and established a formula to estimate the amount of spring back as shown in Fig. 1. In the second step, we followed a similar analysis process to develop the control measures for spring back in the mass production stage and verify our analysis with the production equipment to estimate the effects of the measures.

Table 1. Problem of dimensional precision and engineering level [2][3]

	Problem	Work	Stress condition	Engineering solution level
Local spring back	SB with angle change		Stress difference in cross-sheet direction	◯ Engineering data available
	Wound wall		Stress difference in cross-sheet direction	◯ Engineering data available
Total curvature SB	SB of overall shape	SB	Internal stress of sheet	△ Engineering based on experience
	Cross sectional torsion	Torsion	Internal stress of sheet	△ Engineering based on experience

Table 2. Current engineering level and problems

	Method	Current engineering level	Calculation workload	Accuracy
①	Empirical know-how	· Based on data from the past production and taking shape change into account →Estimate the amount of SB and incorporate into die design (likely to reengineer the die after the trial)	○	✕
②	Elasto-plastic dynamics	· Combination of factors: stress in cross-sheet direction, sheet internal stress and shape factor →Difficulty of analytical solution	✕	△
③	FEM	· Analysis with three dimensional elasto-plastic FEM is necessary →High workload for preparation and calculation delayed the analysis attempt (development expected)	✕	◌
④	Experimental induction (empirical formula)	· Individual part data available · Accumulation of quantitative engineering data not started	○	△

Focus of this study ○

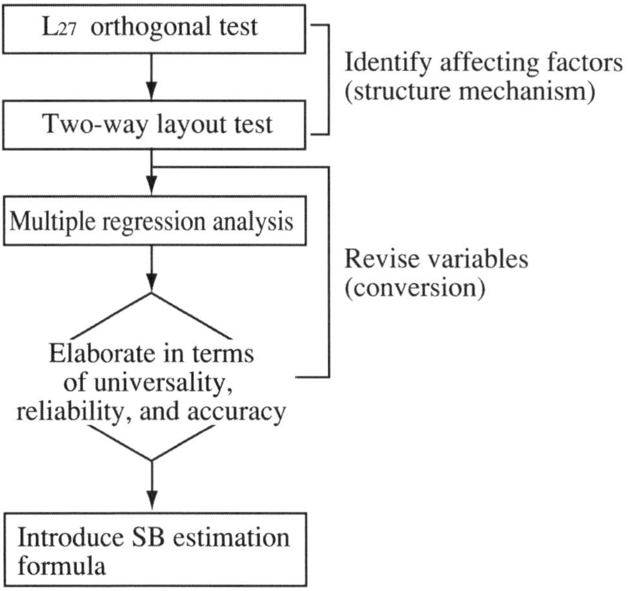

Fig. 1. Analysis procedures, Step 1

2. Study of factors and conditions of experiments

In order to identify the factors, we plotted a systematic chart of the factors (chart omitted) for the thorough engineering study and finalized the factors and standards for the study as shown in Table 3. Some factors that need to be considered are die depth, bearing height and yield point. Die depth and bearing height are the determinants of the product rigidity and restrain spring back while the yield point of the material controls the residual stress that is a cause of spring back.

In addition, we also took die radius, binder surface pressure, and end shapes of the product into account as these factors are supposed to be related to spring back from our experience. Some other factors that may affect the occurrence of spring back are product thickness and width. However, in this study that uses the test plant, we used the center values of the practical range as the fixed figures for the thickness and width because these factors are determined by the product specification.

Table 3. Factors and standards

Symbols	Factors (Standards)	1	2	3
R	Die R (mm)	1,200	5,000	5,000
h	Bearing height (mm)	0	3	5
H	Die depth (mm)	15	20	25
Yp	Yield point (kg/mm2)	18	21	34
C	Binder surface pressure (t)	0	25	50
F	Shape of both ends	Open	Closed	Closed

Sheet thickness: 1.0 mm
Work width: Fixed at 80 mm

In this study, the definition of spring back amount is as follows: The amount of spring back is the difference ($\Delta\theta$) between the central angle of the die radii (θ) and the central angle of the panel radii ($\theta 1$) in the span of 200 mm as shown in Fig. 2.

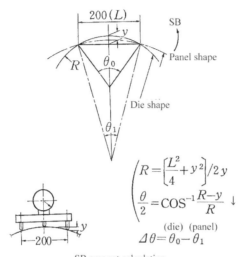

$$R = \left[\frac{L^2}{4} + y^2\right]/2y$$

$$\frac{\theta}{2} = \cos^{-1}\frac{R-y}{R}$$

$$\text{(die)} \quad \text{(panel)}$$
$$\Delta\theta = \theta_0 - \theta_1$$

SB amount calculation

Fig. 2. Definition of spring back

3. Results of analysis

(1) L27 orthogonal test: We conducted an L27 orthogonal test to examine the contribution of each factor to spring back. The results are shown in Fig. 3 (a). We further analyzed the relation between each factor and spring back and found that the results of Yp, H, R and h were consistent with the former studies.

On the other hand, the test showed the opposite results for the end shapes F. Before the test, we estimated less spring back for the closed shape than for the open shape because the stress difference in the cross-sheet direction at the upper die contact area becomes smaller. However, we obtained larger $\Delta\theta$ for the closed end shape as shown in Fig. 3 (b).

(2) Two-way layout test: In order to further analyze the influence of the end shape, we conducted a two-way layout test. The test demonstrated the same

results as shown in Fig. 3 (b) at every yield point (as shown in Fig. 3 (c)). We measured the strain in the lateral direction (ε_A, ε_B) and found that not only the stress at the contact area with the upper die, but the gap between the stress at the contact area and that of the flanged portion (strain, in other words) may affect $\Delta\theta$ as shown in Fig. 3 (d).

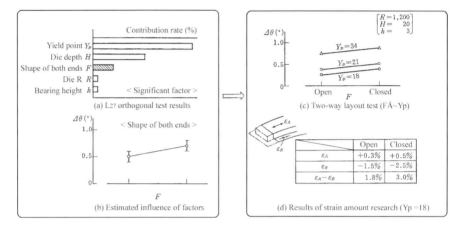

Fig. 3. Results of analysis

4. Results of multiple regression analysis

(1) First analysis: We conducted a linear analysis based on the data collected from the production equipment for prototype as shown in Fig. 4 (a). In this case, the residual increased according to the amount of spring back and curve relation was observed. In order to confirm the stability of the analysis results, we made back-estimation of the L27 orthogonal test results with the multiple regression formula, plotted the estimated results with x in the same figure and found the estimation was almost consistent with the actual L27 orthogonal test results.

(2) Second analysis: In order to increase the accuracy of the estimation formula, we converted data (1) with the data obtained from L27 orthogonal test.
All variables were log-converted. The analysis results are shown in Fig. 4 (b).
After we stratified the results into open and closed shapes, we noticed a slight

$$\Delta\theta \rightarrow \Delta\theta / \theta_o \cdots (1)$$

difference in the slope of the lines. We compared the standard partial regression coefficients of open and closed shape cases and confirmed the difference in the effect of the strain factors (R, h) as shown in Fig. 5.

(3) Third analysis: Based on the results of the second analysis, we analyzed the stratified data of open and closed shapes and obtained the estimation formula that have practical contribution and estimated accuracy in the both cases as shown in the Fig. 6 (a) and (b).

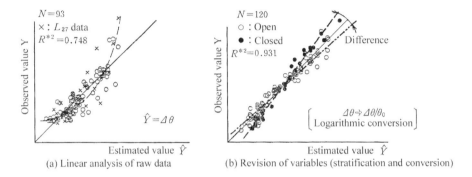

Fig. 4. Results of multiple regression analysis

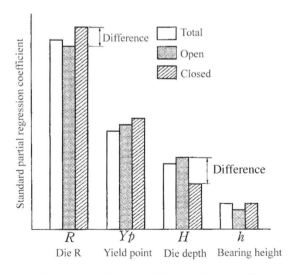

Fig. 5. Comparison of standard partial regression coefficients of factors

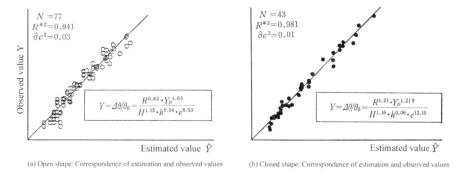

(a) Open shape: Correspondence of estimation and observed values   (b) Closed shape: Correspondence of estimation and observed values

Fig. 6. Results of analysis

5. Study

We looked at the residual stress of the panel after the plastic deformation as shown in Fig. 7. Our test showed that when the panel is stamped, the contact area with the upper die holds positive residual stress while the flange portion holds negative residual stress. This means that after the stamping pressure is removed, shrinking force is generated in the contact area with the upper die and elongation force is generated in the flange portion. As a result, elastic recovery occurs in the direction that increases the product R. Therefore, we can assume that maintaining the balance between the stress in the contact area with the upper die and that in the flanged portion will control spring back.

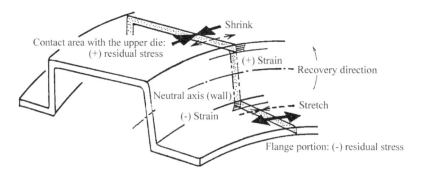

Fig. 7. Analysis of residual stress of panel after plastic deformation

6. Examine control measures for spring back

In many cases, incorporation of the spring back control into the stamping die is difficult when the part shape is complex. Technology to control spring back is the biggest challenge to attain high precision of parts. We plotted a systematic chart of control technologies for the comprehensive evaluation of the feasibility and found the emboss restrike method to be prospective as shown in Fig. 8 (a). From the figure, this method is applied in the process for some parts, but the technology has yet to be theoretically proved as the control measures.

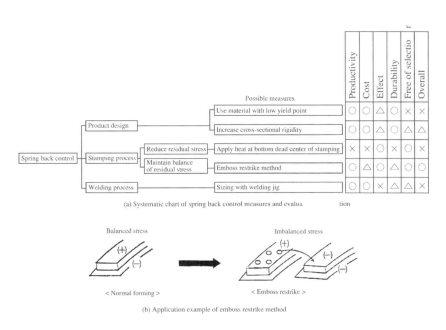

(a) Systematic chart of spring back control measures and evaluation

(b) Application example of emboss restrike method

Fig. 8. Examination of spring back control measures

7. Analyze emboss restrike method

As the second step of the analysis, we conducted an experimental study with steps similar to the first step in order to quantitatively analyze the effect of the emboss restrike method. We omit describing the analysis results, but the standards and factors selected for L27 orthogonal test are shown in the Table 4.

Fig. 9 (a) shows the effects of the major factors on $\Delta\theta$ in the two-way layout test that we carried out for confirmation purposes.

From the analysis, we found some cases in which $\Delta\theta$ became negative (spring forward) as opposed to the case where $\Delta\theta$ became positive (elastic recovery) depending on the combination of the significant factor E (emboss pitch) and the standards such as Yp (yield point). Also, from the analysis shown in the figure, we found that proper emboss alignment minimizes the amount of spring back.

This means that the imbalanced stress as well as the residual stress (positive and negative) at the contact area with the upper die and the flanged portion can be minimized and controlled by the emboss restrike method. For further analysis, we studied the change in the residual stress (from positive to negative) and the spread of the affected area along with the change as shown in Fig. 9 (b). We

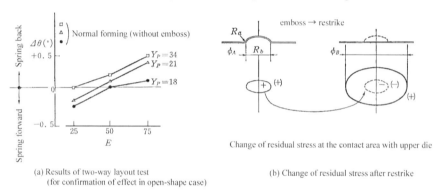

(a) Results of two-way layout test
(for confirmation of effect in open-shape case)

(b) Change of residual stress after restrike

Fig. 9. Analysis of emboss restrike method

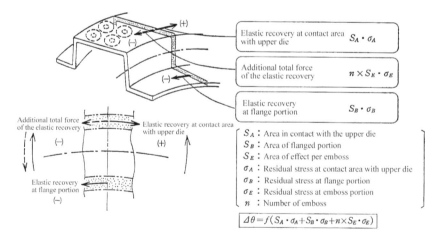

Fig. 10. Model of internal stress balance

then established a model of the internal stress balance.

Force that is generated at the panel is largely classified into the force of elastic recovery at the contact area with the upper die and the flange portion and the additional total force of the elastic recovery at the emboss. The former increases the amount of spring back and the latter works to decrease the amount as shown in Fig. 10. The relation between these two types of force and $\Delta \theta$ is as in formula (2)

$$\Delta \theta = f(SA + \sigma A + SB + \sigma B + n \cdot SE \cdot \sigma E) \quad \cdots \quad (2)$$

Based on formula (2) and the test results, we conducted multiple regression analysis to establish a formula to estimate $\Delta \theta$ as shown in Fig. 11 (a) and Fig. 11 (b). Taking into account the behavior of both spring back and spring forward ($\Delta \theta < 0$), we converted the data as in (3). We made data conversion (4) and logarithm conversion of all variables in order to increase additivity.

$$\Delta \theta \rightarrow 1 - \Delta \theta / \theta_0 \quad \cdots \quad (3)$$
$$E \rightarrow n \cdot SE / (SA + SB) \quad \cdots \quad (4)$$

From the Fig. 11, we could derive the estimation formula for open and closed end shapes that has a practically sufficient contribution rate and assumed accuracy.

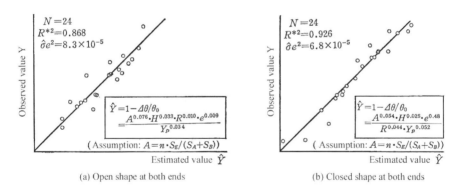

(a) Open shape at both ends   (b) Closed shape at both ends

Fig. 11. Results of analysis

8. Confirm the effect of emboss restrike

We selected actual parts that showed a large amount of spring back in the first trial. We then applied the estimation formula to find the required number of emboss and designed the stamping die accordingly. With this single control measure, we could obtain the required parts precision as shown in Fig. 12.

9. Conclusion

(1) We clarified the effects of shape and forming factors on total curvature spring back.

(2) We derived the estimation formula to enable die design that incorporates spring back assumption.

(3) We selected the emboss restrike method as a spring back control measure, systematically analyzed the relation between the effect range of the emboss and spring back, and devised an estimation formula to control spring back.

These quantitative findings will serve as supportive information for die engineering that enables spring back control at the die design stage.

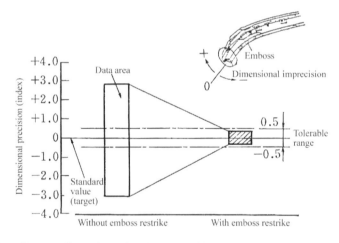

Fig. 12. Effect confirmation of emboss restrike method (e.g. rail, roof side inner)

10. Supplements

This document describes the application of SQC to assist CAE and CAD/CAM and its effect. With the SQC method, we successfully made a systematic analysis of the relation between the amount and direction of residual stress and the rigidity of the work in the following three steps.

Firstly, with the design of experiment, we clarified the shape factors and forming factors that may affect spring back. Secondly, we analyzed the mechanism of spring back using a simple model, stratified the data, and conducted multiple regression analysis to derive the theoretical estimation formula. Thirdly, combining the results obtained with the previous two SQC methods, we established a spring back control technique for stamping.

Recently, Toyota Motor Corporation announced similar studies[4]-[13] in which the physiochemical approach and scientific insight were the key to the findings. To this end, the "Science SQC" approach is promoted as a strong methodology of scientific analysis in the company.

References

[1] K. Kusune, Y. Suzuki, S. Nishimura and K. Amasaka, (1992), "The Statistical Analysis of the Spring Back in Stamped Parts with Large Curvature", (in Japanese) *Quality, Journal of the Japanese Society for Quality Control,* 22(4), 24-30.

[2] "Handbook: Difficulty in Stamping", (in Japanese) *Thin Plate Forming Engineering Institute* (1987).

[3] "Examples of Dimensional Precision in Stamped Vehicle Panels", (in Japanese) *Thin Plate Forming Engineering Institute* (1984).

[4] N. Suzuki, (1990), "Analysis of Press Die Edge Precision in Case Burr is Produced", (in Japanese) *Quality Control, Union of Japanese Scientists and Engineers,* 41(11), 458-463.

[5] S. Mekata, (1991), "Estimation of Contraction in Bumper", (in Japanese) *Union of Japanese Scientists and Engineers, 14th Multivariate Analysis Symposium,* 21-26.

[6] N. Kurume et al., (1991), "Measures for Shrinkage and Porosity in Low Pressure Casting", (in Japanese) *Quality Control, Union of Japanese Scientists and Engineers,* 42 (11), 394-399.

[7] F. Komuro and M. Ito,(1991), "Proposal for an Evaluation Method for Ease of Product Assembly", (in Japanese) *The Japanese Society for Quality Control, 39th Technical Conference,*49-52.

[8] K. Keishima and M. Okada, (1992), "Measures To Be Taken for Difficulties in the Manufacture of Aluminum Die-cast Head Covers", (in Japanese) *Total*

Quality Control, Union of Japanese Scientists and Engineers, 43(5), 86-91

[9] T. Hayashi and T. Kido,(1992), "Towards a Pleasant Working Environment", (in Japanese) *The Japanese Society for Quality Control, 22nd Annual Technical Conference,*65-68.

[10] H. Nishimura, (1992), "A study on Making Automotive Internal Materials Lighter", (in Japanese) *Union of Japanese Scientists and Engineers, 16th Multivariate Analysis Symposium,* 81-86.

[11] A. Takeuchi et al., (1992), "A Study on the Appearance of Aluminum Wheels", (in Japanese) *Union of Japanese Scientists and Engineers, 16th Multivariate Analysis Symposium,* 9-14.

[12] S. Kameyama et al., (1992), "A Study on the Method of Visual Inspection", (in Japanese) *The Japanese Society for Quality Control, 42nd Technical Conference,*57-60.

[13] K. Hayashi,(1992), " Improvement of Painting Color Matching for Vehicles", (in Japanese) *Quality Control, Union of Japanese Scientists and Engineers,* 43 (5), 194-198.

Chapter 14: Process Design

Improvement of Center-less Ground Surface Roughness

Studies on the Rust Preventive Quality Assurance
of Rod Piston Plating Parts

This study is characterized not only by a measure for improving corrosion prevention through improvement of plating method but also by a new machining process devised to reduce the production process and the running cost. The core technology is applied to improving grinding method. It adopts sealing technique that crushes the opening of cracks to subsequently achieve remarkable improvement of Q, C and D.

In the present research, multivariate analysis method that has been established as a "Science SQC" approach is effectively used by the engineering staff for their dexterous execution of engineering studies.

Keywords; Rod Piston Plating Parts,Rust Preventive Quality Assurance, Center-less Ground Surface Roughness, "Scuence SQC".

1. Introduction

In recent years, this company has been positive in applying "Science SQC" as a powerful weapon of technical analysis to solve new technologies, new methods and other engineering tasks[1-5]. This report describes part of the application through "QCD research activities under a new method" with the cooperation of engineering staff from the production engineering and manufacturing divisions. It takes up the improvement of corrosion resistance of rod piston, which is a principal component, with the objectives of improving rust prevention performance of shock absorber that determines ride quality of a vehicle.

The sliding portion of a rod piston is chrome-plated as shown in Fig. 1 to improve corrosion resistance. Cracks are generated, however, in the direction of

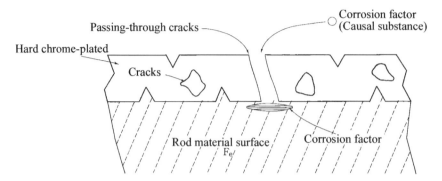

Fig.1. The Mechanism of the Rust Occurrence

depth by internal stress, which is generated during the formation of plating film. If the cracks are large, they are mutually connected to form a passing through crack. The presence of passing through crack allows corrosion factor (causal substance) to infiltrate the rod material surfaces to cause rust.

The present research is characterized not only by a measure for improving corrosion prevention through improvement of plating method but also by a new machining process devised to reduce the production process and the running cost. The core technology is applied to improving grinding method. It adopts sealing technique that crushes the opening of cracks to subsequently achieve remarkable improvement of Q, C and D (Quality, Cost and Delivery).

In the present research, multivariate analysis method that has been established as a "Science SQC" approach is effectively used by the engineering staff for their dexterous execution of engineering studies[6-7].

2. Improvement of Corrosion Prevention Performance

The present studies first began with the elimination of passing-through cracks. To do this, we adopted two approaches, namely (a) development of thicker plating film, and (b) development of micro-crack (change of plating assistant). As the result, generation of passing-through cracks could be drastically prevented but not completely. Next, a new sealing method was studied by providing parts completed with plating with super finishing to form affected layer on the surface of the plated film, (c) thus crushing cracks.

Characteristics of the present studies are not only adding super finishing but also

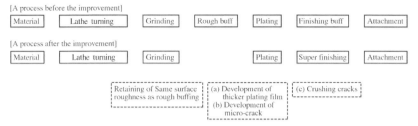

Fig. 2. The Process Flow Diagram

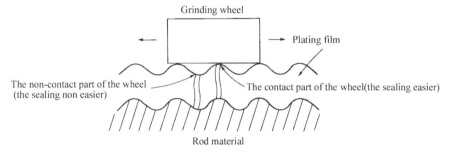

Fig. 3. Relationship between the formation by the Plating film and the Sealing by the Super finishing

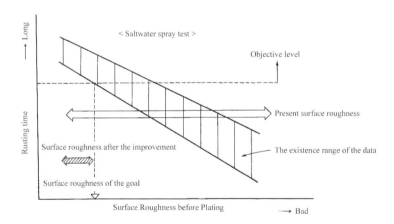

Fig. 4. Relationship between Surface Roughness before Plating and Rusting time for Corrosion Prevention Performance

simplifying the process and improving corrosion prevention quality compatibly by improving respective process to keep the cost low as shown in the process flow diagram of Fig. 2. One of the contents of improvement to make first was to switch grinding wheel quality from a general grinding wheel using binding material of vitrified system to UB grinding wheel using elastic fiber grinding wheel so as to retain same surface roughness as rough buffing . Secondly, it was to improve the plating condition to make the sealing easier with the super finishing. These improvements eliminate the rough buffing and finishing buff processes simultaneously.

However, if the base of rod material on which plating film is formed is irregularly surfaced, subsequently formed plating film will trace the contour of rod material as shown in Fig. 3. It has been observed that this prevents the super finishing wheel from making uniform contact with the plating film surface, producing incompletely sealed through crack as the result.

In addition, it has been discovered from the above that there is a technically significant corresponding relationship between surface roughness before plating and corrosion prevention performance (rusting time is used for the index here). That is, the worse the surface roughness, the lower corrosion prevention performance drops as discovered from a saltwater spray test shown in Fig. 4. In this connection, to meet the objective level, it was necessary to drastically improve the surface roughness of rod material before plating.

The present studies have taken up centerless grinding machine and grinding as the means for improving surface roughness of rod material. Prescribed engineering findings and results are obtained by applying scientific SQC approach, which are reported as follows.

3. Examination and Boil-down of Factors

3.1. Centerless Grinding Machine and Grinding

Centerless grinding is performed by holding the work by whole surface, which keeps the work under machining with little deflection, hence suited for mass production. Fig. 5 shows the representative through-feed grinding of centerless grinding machine. A long, cylindrical work is positioned between the grinding wheel and the adjust wheel. It is held loosely on a V-shaped surface configured with the peripheral surface of the adjust wheel and the top face of the blade (backing plate).

The work is ground by the wheel while being driven to turn with the grinding force. Braked by the friction force generated with the adjust wheel, it turns slowly at the rate of peripheral velocity of the adjust wheel independently from the peripheral velocity of the grinding wheel. At this time, the work, held by the supporting surfaces, is ground to become circular while being aligned continually. Moreover, as the work is turned and ground, it is fed simultaneously in the direction of an arrow in Fig. 5 along the guide plate by the peripheral velocity and the incline angle (feed angle) of the adjust wheel.

Studies concerning the accuracy of grinding with the application of SQC comprise, for example, a multi-pin grinding machine for engine crankshaft[8-10] and CNC grinding[11]. Few cases of studies seek causal relation quantitatively concerning the accuracy of machined surface (hereinafter referred to as surface roughness) provided by the centerless grinding machine, however.

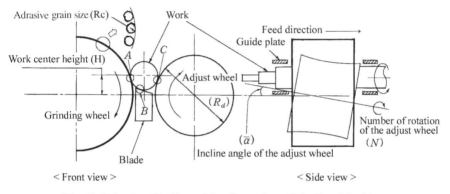

Fig. 5. Grinding Outline of the Centerless Grinding Machine

3.2. Examination and Boil-down of Factors through Technical Investigation

To improve surface roughness before plating, it is necessary to have the surface roughness built in during the grinding process. For example, Waguri[12] et al. simplified the theoretical surface roughness (S) of the lathe-machined surface as $S = V2/8R$ where (R) is the roundness of the tip of a cutting tool and (V) the feed velocity. Actual roughness of ground surface, however, is generally similar but it becomes greater due to various reasons such as the grinding mechanism, condition and state of grinding, quality of cutting tool, vibration during machining, etc. Sometimes it varies greatly and complicatedly.

On the other hand, the grinding theory leads the grinding operation by assuming that the abrasive grains of the wheel work the same way as the milling edge. Reality is more complicated and the theoretical roughness of the ground surface is not necessarily simplified. Conversely, reality and theory are often quite different from each other. In this sense, not much clarification has been made of quantitative causal relation concerning the roughness of ground surface.

Generally speaking, most of studies on important centerless grinding technique concern circularity-forming work of how to make a distorted work as circular as possible by holding it with points B and C and grinding at point A as shown in Fig. 5. In many cases, required roughness of ground surface for machined parts is achieved within the scope of those studies.

However, to establish the sealing technique as being discussed in this study, a rank higher surface roughness is required by fully suppressing irregularity on the ground surface. While, explanatory factors that need to be controlled to obtain required surface roughness are not necessarily specified nor quantitative effect is clarified since these techniques are empirical and qualitative.

In this connection, upon making examination of factors for surface roughness, causal relation of surface roughness has been further re-examined technically from both aspects of cutting and grinding theories on the basis of empirical technique.

As the result, by assuming that the ground surface roughness (Sc) depends on the work through-feed velocity (Vf) and the grain size of wheel (Rc) and that the work through-feed velocity (Vf) can be specified by the diameter of adjust wheel (Rd), the incline angle (α) and the number of rotation (Nr), the following equation has been introduced from the viewpoint of a proprietary engineering technology:

$$Sc = Vf^2 / C \cdot Rc \quad \cdots\cdots\cdots\cdots\cdots (1)$$
$$Vf = \pi \cdot Rd \cdot \sin \alpha \cdot Nr \cdots\cdots (2)$$

Next, centerless grinding machine and the work are arranged to factorial systematic diagram, etc. from the viewpoint of proprietary engineering technology. Then, to improve surface roughness, the importance are examined and reviewed on the "degree of factorial effect and the "freedom of selection" of factors and levels for control.

On the basis of the result, it was first decided to acquire the effectiveness as an empirical technology. Here, four explanatory factors were adopted to examine their causal relations. They are the "number of rotation" and the "incline angle " of the adjust wheel of grinding machine, and the "work center height" and the "grinding allowance", which are important in securing circularity during grinding.

4. Examination and Analysis of Causal Relations by Multivariate Analysis

Upon examining and analyzing causal relations, it is difficult to carry on tests by stopping the equipment in operation on a production line. Therefore, the following analyses were made using Toyota multivariate analysis software (TPOS-PM)[13] and the data outlined in Table 1, which was already sampled during the stage for obtaining optimal condition for mass production on a centerless grinding machine that used UB wheel.

4.1. Preliminary Examination based on Class I Quantification Method

Regarding the explanatory factors in Table 1, it is not necessarily true that there exist linear relations within the scope of available data. To make a technologically significant multiple regression analysis in section 4.2, it is important to understand the effect of explanatory factors qualitatively as a preliminary examination and add discussion to it from the viewpoint of proprietary engineering technology.

In this connection, the four explanatory variables ($X(1)$-$X(4)$) were developed into four items (A to D), then classified into 3 to 4 categories by giving consideration to whether the factors are in linear or non-linear relation or if they have extreme values. Analysis was then made using Class I quantification method.

According to the analytical result of Table 2, effect may exist generally in the following sequence of $X(1)$ incline angle > $X(2)$ number of rotation > $X(3)$ center height > $X(4)$ grinding allowance from the range and partial correlation coefficient although category frequency is not necessarily well-balanced. Then, the causal relation was estimated from the point of engineering technology within an area including the range of existence of the analytical data in the light of

$(N=27)$

Sign	Influence factor	Data
$X(1)$	(α) Incline angle(degree)	$0.8 \sim 3.0$
$X(2)$	(Nr) Number of rotation(rpm)	$27 \sim 45$
$X(3)$	(H) Work enter height (mm)	$10 \sim 12$
$X(4)$	(l) Grinding allowance (μm)	$10 \sim 40$
Y	(S) $(*1)$ Surface roughness (μRZ)	0.175 ~ 0.445

$(*1$ JIS B 601$)$

Table 1. Analytic use data

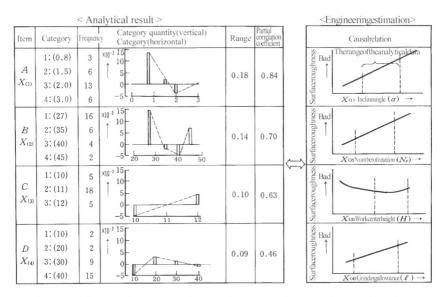

Table 2. The Analytical Result Based on Class I Quantification Method

empirical technology. The analytical result of Class I quantification method and the result of engineering estimation were compared mutually in Table 2. The result indicated a major difference of effect is present with respective factors from the analytical result of Class I quantification method.

This suggests possible interference with other influence factors whose effect cannot be disregarded from the viewpoint of empirical technology. Thus the existence of other important factors, which were not incorporated to the analysis, is assumed. Therefore, to review the causal relation quantitatively in a deeper analysis, another analysis was made as described in the following section.

4.2. Multiple Regression Analysis (1)

On the basis of the analytical result of Class I quantification method of section 4.1, metrological data of the four explanatory factors $X(1)$ to $X(4)$ were incorporated, and first to understand general effect of the four variables, multiple regression analysis (1) was attempted using the variable change method (Fin, Fout; 2.0) (the indication of the analytical result is standardized here. In addition, analytical result is verified by using variable designation method in the following).

As the result, incline angle $X(1)$ and center height $X(3)$ were incorporated from Table 3, which is apparent from the primary effect as analyzed in Table 2. On the other hand, it is understandable that number of rotation $X(2)$ and grinding allowance $X(4)$ were not incorporated because the secondary effect is strong and that the range is small with weak primary effect. R^2 is as low as 0.351 (R^{*2} being 0.299), however, and regression residual has a trait (dotted line) as indicated by

Table 3. The Analytical Result of Multiple Regression Analysis (1)

Multiple correlation	.592539	Adjusted R-square	.299191
R-square	.351103	Akaike's information criterion, (AIC)	−76.387

——— Analysis of variance table ———

(7) Degrees of freedom (8) Sum of square (9) Deviation of Sum of square (10) F-value

	(7)	(8)	(9)	(10)
Regression	2	0.04	0.0200	6.7635 ∗∗
Residual	25	0.08	0.0032	
Total	27	0.12		

NO.	Standard partial regression coefficient, ($\acute{E}_i$-weight)	Partial correlation coefficient	Standard error of standard partial regression coefficient, (Standard error of $\acute{E}_i$-weight)	F	VIF
$X(1)$	−0.4899	−0.51949	0.16114	9.24055 ∗∗	1.0005
$X(3)$	0.3232	0.37231	0.16114	4.02311	1.0005

the scatter of estimated values and measured values in Fig. 6.

The result is apparently insufficient from the reliability of analysis and is unreasonable similar to the analytical result of Class I quantification method. In this connection, further review of explanatory variables has become necessary by reviewing the scatter diagram matrix and making technological interpretation.

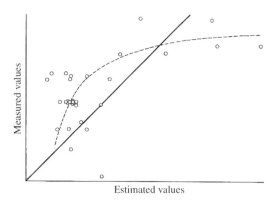

Fig. 6. The Corresponding Scatter Diagram of the Estimated and Measured
    Values (Multiple Regression Analysis (1))

4.3. Review of Explanatory Variables

When the scatter diagram matrix is reviewed concerning the center height $X(3)$ for example, curved relations as indicated in Fig. 7 can be generally surmised and the optimal value (Ho) is thought to be present in the middle of the analytical data range although the data has some dispersion. According to empirical technology, the outer surface of the work becomes pitched easily if the center is too high or circularity becomes deteriorated if too low[14]. Consequently, it is important to quantitatively verify an optimal center height at which favorable surface property can be obtained.

To acquire the optimal center height (Ho) at which required surface roughness can be obtained, secondary item $X(8)$ [$(X(3)-Ho)^2$] is added as the analytical method in addition to the primary item ($X(3)$). The optimal center height (Ho) has been surmised from the figure, and further it has been verified through an additional test on actual machine that the optimal center height (Ho') is present near the center of the range of center height at which good surface property can be obtained.

Moreover, to enhance analytical accuracy, other explanatory factors were newly added on the basis of the factorial system diagram. These factors are thought to have technological effect, yet having not been thus far taken up for analysis. They comprise the type of wheel (2 different diameters) and contour factors of adjust wheel (3 types), which have been added respectively as dummy variables $X(5)$, $X(6$ and $X(7)$. Table 4 indicates explanatory variables after the review.

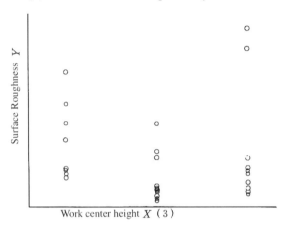

Fig.7. Review of Explanatory Variables-The scatter diagram matrix between the work Center Height and Surface Roughness

Table 4. Review of Explanatory Variables

$X(1)$	Incline angle
$X(2)$	Number of rotation
$X(3)$	Work center height
$X(4)$	Grinding allowance
$X(5)$	Grinding wheel type (two kinds)
$X(6)$	Adjust wheel form (three kinds)
$X(7)$	
$X(8)$	$(X(3) - H'_o)^2$

4.4. Multiple Regression Analysis (2)

As the result of another multiple regression analysis(2) conducted by incorporating those variables, R^2 has been improved to 0.807 (R^{*2} being 0.763) as Table 5 indicates. In addition, no remarkable trait is observed about the regression residual as the corresponding scatter diagram of the estimated and measured values shows in Fig. 8. Error variance from regression is about 1/4 of multiple regression analysis (1), which is practically small and on an appropriate level. The effect of explanatory variables incorporated from the table was $X(7) > X(3) > X(5) > X(8) \doteqdot X(6)$ in descending order in terms of standard partial regression coefficient.

Form factors of the wheel type $X(5)$, adjust wheel form $X(7)$, the primary ($X(3)$) and the secondary ($X(8)$) of center height are important controllable explanatory factors, which respectively have effect on the grinding power of work surface and the rate of work rotation and through-feed velocity. The result of analysis is technically reasonable. Moreover, regarding the center height, Table 5 indicates that the standard partial regression coefficient of the primary ($X(3)$) is greater than the secondary ($X(8)$).

In this connection, to double-check the level of influence of the primary effect of center height, another multiple regression analysis was conducted without explanatory variable $X(3)$. As the result, R^2 became 0.693 (R^{*2} being 0.655). Contributory rate was lowered by 11%. It has been verified that the primary effect of the center height has influence similar to or over the secondary effect. This has been checked with the additional test using an actual machine as stated in section

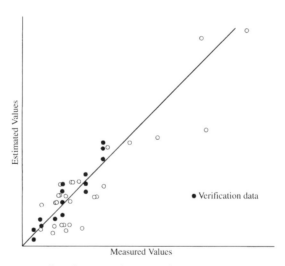

Fig. 8. The Corresponding Scatter Diagram of the Estimated and Measured
    Values (Multiple Regression Analysis (2))

4.3. It also matches the related investigation result on the center height and surface roughness and proves understandable from the view point of technology.

Next, analytical result of Table 5 will be discussed in more detail. From the scatter diagram matrix, there is no strong correlation between the explanatory factors, and +/- codes of respective explanatory variables and the size of standard partial regression coefficient is agreeable from the viewpoint of technology. VIF values for example are between 1.04 and 1.27 as the table indicates, which represent fully valuable result.

Standard error of standard partial regression coefficient is relatively large, however. To control process with clearly defined target, the standard error of standard partial regression coefficient should be reduced to 1/5 or lower of the standard partial regression coefficient. To do this, the analytical accuracy should be raised by adding analytical data or through a factorial positioning experiment using a table of orthogonal arrays, to which new factors are incorporated intentionally.

On the other hand, new findings have been obtained as a control factor although within the scope of test data. These include the effect of the type of wheel $X(5)$, which was much more effective than initially expected, and the size of influence from the form of adjust wheel $X(6)$ and $X(7)$. Quantitative verification will be further advanced on these points.

It has been quantitatively verified presently that incline angle $X(1)$, number of rotation $X(2)$ and grinding allowance $X(4)$ are not much effective within the range of the current test data. This means an expansion of the range of level selection for the control factor, which is significant from the aspect of improvement of Q, C and D.

From the above-mentioned results, equation (3) (standardization) has been obtained as a rational empirical formula that expresses the initial objective of causal relation.

$$Y' = 0.866X(7) - 0.165X(6) - 0.232X(5) + 0.327X(3) + 0.180(X(3) - H'\text{o})^2 \cdots (3)$$

As the optimal conditions, solution of quadratic equation is given to "center height" $X(3)$ and it was assumed that $X(5) = X(6) = 1$, $X(7) = 0$. Regarding the explanatory variables, which were not incorporated, technological viewpoint and the restriction on the production speed (sec/work) were determined.

Table 5. The Analytical Result of Multiple Regression Analysis (2)

| Multiple correlation | .898207 | Adjusted R-square | .762861 |
| R-square | .806776 | Akaike's information criterion, (AIC) | −104.307 |

——— Analysis of variance table ———

	Degrees of freedom	Sum of square	Deviation of Sum of square	F-value
Regression	5	0.10	0.0200	18.3714 * *
Residual	22	0.02	0.0009	
Total	27	0.12		

NO.	Standard partial regression coefficient, ($\acute{E}_i$-weight)	Partial correlation coefficient	Standard error of standard partial regression coefficient, (Standard error of $\acute{E}_i$-weight)	F	VIF
$X(7)$	0.8659	0.88065	0.09932	76.01130 * *	1.1231
$X(5)$	−0.2921	−0.50894	0.10432	7.69053 *	1.2630
$X(3)$	0.3273	0.59014	0.09547	11.75590 * *	1.0377
$X(8)$	0.1803	0.34238	0.10551	2.92130	1.2675
$X(6)$	−0.1650	−0.33776	0.09803	2.83291	1.0942

4.5. Implementation of Verification

Fig. 8 shows the result of grinding performed for respective values of explanatory variables with small black bullet holes (·). Their values are changed by combining several types of explanatory variables technically by observing the vicinity of required surface roughness. As the result, it has been verified that the model equation (3) has good applicability.

5. Summarization

5.1. Implementation and Achievement

The process flowchart shown in Fig. 2 represents a vertical flow process under the so called TOYOTA Production System. The production process is configured with multiple equipment. As the result of lateral development of optimal conditions obtained from the analysis thus far described to the production equipment, mechanical error between the equipment has been eliminated. This drastically improved the surface roughness of works before plating as indicated with the arrow in Fig. 4, thus holding irregular heights on the ground surface fully under control.

As the result, sealing technique before and after super finishing of plated surface layer of the work has been established as Photo 1 shows. The original objectives of (1) corrosion prevention performance of rod piston has become 4 times compared with before the improvement. Thus the target of three times as much corrosion prevention performance has been achieved without adding new equipment.

In addition to the above, the present result of analysis led to a discontinuation of rough buffing and replacement of the finish buffing and super grinding machine. This has greatly contributed to the cost reduction including (2) reduction of process (reduction of lead time) and (3) a 15% reduction of running cost. Moreover, quantitative acquisition of control factors for obtaining an optimal surface property using the centerless grinding machine enabled standardization for quality control.

5.2. Accumulation of Technologies

One of the production technologies presently accumulated through the QCD research activities is an establishment of sealing technique after plating. Another is an acquisition and verification of an empirical formula that technologically predicts and controls the required roughness of grinding surface before plating for super grinding.

Photo 1. Sealing after Super Finishing of Plated Surface Layer of Work

5.3. Future Research Tasks

Every alignment of a centerless grinding machine is said to affect each other mutually[12]. The present studies could achieve the target smoothly by analyzing thinkable principal factors in the utilization of sampled data. It is intended to further promote physicochemical approach to the subject by acquiring the effect from other factors quantitatively for discussion, simultaneously using experimental design to prevent the cost of tests from rising.

6. Postscript

Recently, Toyota Motor Corporation announced similar studies[8, 15-19] in which the physiochemical approach and scientific insight were the key to the findings. To this end, the "Science SQC" approach is promoted as a strong methodology of scientific analysis in the company. Company-wide SQC promotion activities this company has thus far advanced are steadily expanding to among Toyota group companies[20-22].

References

[1] K. Amasaka, "Application of Classification and Related Methods to the SQC Renaissance in Toyota Motor", *Data Science ,Classification and Related Methods,* 684-695,(1998), *Springer.*

[2] K. Amasaka, "A study on "Science SQC" by Utilizing "Management SQC" - A Demonstrative Study on a New SQC Concept and Procedure in the Manufacturing Industry-", *Journal of Production Economics,* 60-61, 591-598,(1999),*Elsevier.*

[3] K. Amasaka and S.Osaki, "The Promotion of New Statistical Quality Control Internal Education in Toyota Motor - A Proposal of "Science SQC" for Improving the Principle of TQM-", *European Journal of Engineering Education (EJEE)* ,24(3), 259-276, (1999).

[4] K. Amasaka, "A Demonstrative Study of a New SQC Concept and Procedure in the Manufacturing Industry-Establishment of a New Technical Method for Conducting Scientific SQC-", *An International Journal of Mathematical &*

Computer modeling, 31(10-12), 1-10 (MAY-JUNE,2000).

[5] K. Amasaka, "Proposal and Implementation of the "Science SQC" Quality Control Principle", *International Journal of Computer Modeling,*38(11-13), 1125-1136, (2002).

[6] K. Amasaka and K.Maki, "Application of Multivariate Analysis for the Attraction of Manufacturing Vehicles", (in Japanese) *The Behavior Metric Society of Japan, The 19th Annual Conference,* 64-69, (1991).

[7] K. Amasaka and T. Kosugi, "Application and Effects of Multivariate Analysis in Toyota", (in Japanese) *The Behavior Metric Society of Japan, The 19th Annual Conference,* 178-183, (1991).

[8] M. Koiwa and K. Amasaka, "Application of Multivariate Analysis to Poor Accuracy Diagnosis - A Method of Establishing Grinder Equipment Diagnosis Technology-", (in Japanese) *The Japanese Society for Quality Control, The 42nd Technical Conference,* 37-40, (1992).

[9] H. Irie, "Improvement of Crankshaft Journal Grinding Accuracy", (in Japanese) *Quality Control, Union of Japanese Scientists and Engineers,* 39(5), 67-70, (1988).

[10] K.Tabuchi, "Improvement of Crankshaft Pin Grinding Accuracy", (in Japanese) *Quality Control, Union of Japanese Scientists and Engineers,* 41(5), 76-81, (1990).

[11] Y. Oda, "Trial Production and Evaluation of the Grinding Simulator ", (in Japanese) *The Japanese Society for Quality Control, The 37th Technical Conference,* 13-16, (1992).

[12] A. Waguri, "The method of the machining", (in Japanese) Youkendo (1978).

[13] K. Amasaka and K. Maki, "Application of SQC Analysis Software in Toyota", (in Japanese) *Quality, Journal of the Japanese Society for Quality Control,* 22(2), 79-85, (1992).

[14] S. Yonezu, "The Method of Manufacturing Using Centerless Grinding Machine" , (in Japanese) *Nikkan-Kougyou Shinbun-sha,* (1996).

[15] T. Noiri and T. Azukisawa, "Study on Crack Prevention for Sintered Automobile Parts", (in Japanese) *The Chubu Quality Control Convention, Chubu Quality Control Society,* 198-216, (1990).

[16] M. Miyamoto et al., "Establishment of Optimum Casting Conditions for Compact DOHC Cylinder", (in Japanese) *Quality Control, Union of Japanese Scientists and Engineers,* 42 (5), 183-189, (1991).

[17] T. Takasago and K. Tozuka, "Development of Laser Welding Method for Press Materials of Different Thicknesses", (in Japanese) *Quality Control, Union*

of Japanese Scientists and Engineers, 43 (11), 93-97, (1992).

[18] K. Amasaka et al., "Consideration of Efficientical Counter Measure Method for Foundry,-Adaptability of Defects Control to Casting Iron Cylinder Block-", *The Japanese Society for Quality Control, The 47th Technical Conference,* 60-65, (1994).

[19] K. Amasaka et al., "The Development of Working Condition Taking the Lead an Epoch (#1)(#2)", *The Japanese Society for Quality Control, The 57th Technical Conference,* 53-60, (1997).

[20] K. Amasaka and H. Azuma, "The Practice of SQC Training at Toyota-Improving Human Resource Development and Practical Outcomes", (in Japanese) *Quality, Journal of the Japanese Society for Quality Control,* 21(1), 18-25, (1991).

[21] K. Amasaka, "The Development of the SQC Renaissance to Toyota Group", (in Japanese) *Standardization and Quality Control, Japanese Standards Association,* 45(5), 64-58, (1992).

[22] K. Amasaka, (Editing Project Chairperson), "Science SQC-The Reform of the Quality of the Business Process", (in Japanese) *Nagoya QST Society Edition, Japanese Standards Association,* (2000).

Chapter 15: Production Preparation

Improving Equipment Reliability:

Improving the Reliability of Body Assembly Line Equipment

It is necessary to establish higher levels of equipment reliability in a short time, the market demands ever shorter lead times for the release of new models. Also, the demand for new-model cars is very strong immediately after their introduction. The conventional method for enhancing equipment reliability is by only screening. However, this requires screening operations on production lines and so has been an obstacle to line production and prevented shortening of lead times.

We are now able to dramatically enhance equipment reliability in a very short time by detecting failure modes and forecasting the number of occurrences using a scientific technique based on reliability engineering.

Keywords: Weibull Analysis, m (shape parameter), MTBF (mean time between failures), Defect Control Monitor, Nighttime Durability Test.

1. Introduction

In recent years, flexibility and high precision in body production equipment hasbecome a necessity, leading Toyota to introduce FBL (flexible body line) into our plants. FBL uses numerous robots and pallets in a circulating system consisting ofseveral assembly lines, with a main body line and under body line serving as thecore. The assembly line layout is shown in Fig. 1. In the concept of the lean production system being adopted by Toyota (in which there are very few buffers and parts flow one-dimensionally before being assembled), our primary goal is to realize an assembly line that is free of defects.

In the past, the reliability of the equipment was ensured by actualizing the foreseeable problems through durability tests and taking of appropriate measures in order to prevent problems from occurring after the line-off (hereafter L/O). However, these previous techniques alone are no longer adequate to satisfy the demand for increased quality.

Hence, at this time, we have applied reliability technology along with scientific analysis to obtain a method which produces maximum results through minimum efforts (rationalized measure plan). We report on this method in this article.

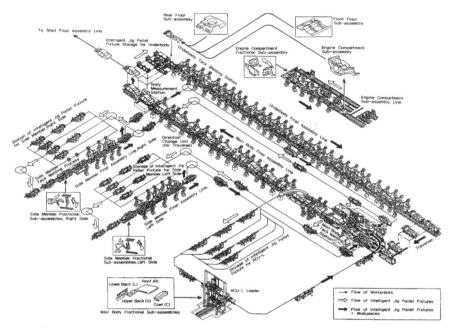

Fig. 1. Overview of final assembly line

2. Background of Present Activity

2.1. The need for improving equipment reliability

Recently, there has been a sharp rise in the demand for new vehicles immediately following a L/O. When this is compared to the performance of the line capacity utilization rate of previous projects, it is evident that they did not attain their goals. This is indicated Fig. 2.

Therefore, ensuring an extremely high level of equipment reliability immediately following L/O has become an absolute technical condition, with past techniques alone not being able to offer satisfactory results.

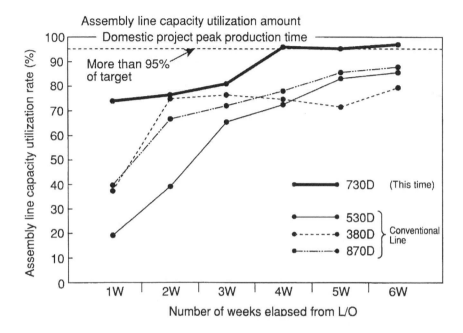

Fig. 2. Recent domestic project assembly line capacity utilization rate

2.2. Current state of body assembly line equipment

The equipment reliability methods of the past aimed to put equipment in operation at an early stage, uncover defects early, and improve the reliability of the equipment. In practice, this meant that between the primary and secondary quality introduction phases of prototype during the process maintenance period, a cycle operation would be performed during a three-night durability test operations as shown in Fig. 3.

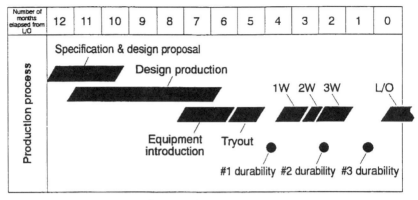

Fig. 3. Example of production equipment operation plan

3. Objectives of Curren Activities

3.1. Problems with past methods

In the past, countermeasures in terms of equipment defects were limited to controlling the quantity of defects. An example is given in Fig. 4. However, recent equipment design, production technology and equipment quality have made spectacular improvements, resulting in a decrease in the quantity of defects and taking the control of the quantity of defects to its limit. New analysis methods are therefore required to reach extremely high target capacity utilization rates and achieve further improvements.

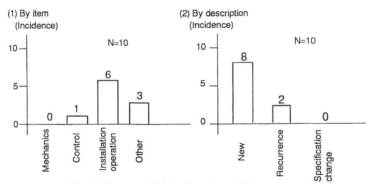

Fig.4. Report of nighttime durability test results

3.2. Application of Weibull Analysis

To pursue the nature of the defects, we decided to perform a scientific failure analysis by applying reliability technology (Weibull Analysis). According to this method, "time" is incorporated as a parameter of the failure data, so that the level of the failure (stop time) and when it occurred (clock time) can be monitored, allowing the analysis of the intervals of malfunction occurrences and definition of the malfunction modes.

Generally speaking, defects can be distinguished as to whether they are initial, sudden, or wear defects according to their "m" (shape parameter) obtained through Weibull Analysis as shown in Fig. 5.

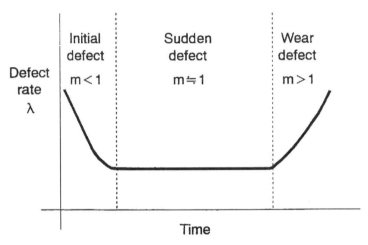

Fig. 5. Bathtub curve line

3.3. Objectives of the present activity

The implementation of Weibull Analysis made it possible to discover the three items below for the first time. Accordingly, it has become possible to plan detailed and effective countermeasures for each defect, from which distinctive equipment reliability can be anticipated.

(1) Forecasting how defects will occur in the future.
　(Making it possible to plan a system and arrangement in terms of equipment.)

(2) Analyzing current equipment capabilities on a quantitative basis.

(3) Consideration of appropriate countermeasure methods in terms of current equipment defects.

4. Results of the Present Activity

4.1. Analysis procedures and their examples

A defect control monitor has been utilized to collect defect data from the actual assembly lines. As described in Fig. 6, the defect control monitor is a system that collects actual defect data from the andon control devices and, through the communication links, obtains defect data at each assembly line in terms of the time function. In analyzing defect information via this defect control monitor, we are able to visually inspect actual defects according to the "on-location, on-item" concept, and verify that our defect findings conform to the information provided by the defect control monitor.

At the Takaoka Plant's Corolla assembly line, we conducted a nighttime durability test (equipment is operated at night without actually producing vehicles to screen for equipment defects) and, for the first time, we were able to collect data which included the defect occurrence time and line stoppage time.

We verified the "conditions for accurate verification" used to evaluate the reliability of production lines. The car body production line includes rest times (hot time and lunch time) and interruptions between the day and night shifts (time between shifts). We conducted a trial Weibull analysis for these test times and interruptions, and verified the "m" values (shape parameter) and MTBF for lines 1) and 2) mentioned below from plan 1) through plan 4) as shown in Fig. 7.

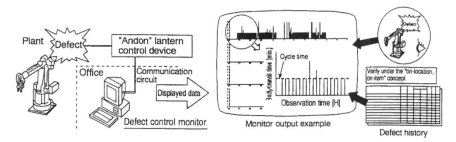

Fig. 6. Defect data collection method

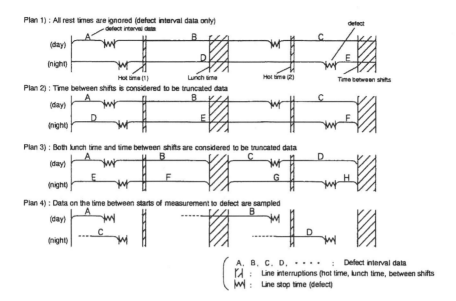

Fig. 7. Sampling defect interval data

1) Lines subject to frequent stops immediately after start of the line : U/B line "initial defect"

2) Lines subject to sudden defects : M/B line "sudden defect" ($m=1$)

In terms of the "m" value, plans 2) and 3) may have a high rate of truncated data, and thus the reliability may be low. Probable causes include the fact that many data are sampled from the defects of the lower ranks. Plan 4) is inappropriate because "m" is greater than 1 ($m>1$). In terms of MTBF, plans 2) and 3) have more truncated data, and smaller values than the actual ones are derived. Plan 4) produces data of essentially different properties. These conditions are shown in Table 1. We therefore eventually adopted plan 1), which is considered to be the most appropriate.

Raw data such as that from the defect history table was then rearranged into defect interval data and input into the Weibull Analysis program for processing. This calculated the respective "m" values (shape parameter) and MTBF as shown in Fig. 8.

Table 1. "m" Values and MTBF by method of Sampling defect interval data

	m value	MTBF (minutes)
Plan 1)	0.69	992.0
Plan 2)	0.47	621.2
Plan 3)	0.75	422.4
Plan 4)	1.15	2912.1

Examples of lines subject to frequent stops immediately after start of line : U/B line (initial defect (m<1))

	m value	MTBF (minutes)
Plan 1)	1.02	345.8
Plan 2)	0.79	242.0
Plan 3)	1.22	213.7
Plan 4)	0.93	3190.6

Examples of lines subject to sudden defects : M/B line (sudden defect (m=1))

4.2. Analysis results

The primary aim of analyzing the second durability test was to determine whether the subject equipment belongs to either category (1) or (2) given below, as classified by their "m" values (shape parameter), and to begin introducing equipment which conforms to their respective equipment conditions.

(1) $m<1$: Future reliability improvement is possible through screening by replacement or adjustment.
(2) $m\geq1$: Requires fundamental design changes.

For the subject equipment, all assembly lines were further subdivided into 11 lines by the mode of transport. The "m" values and MTBF from the results for all 11 lines were then calculated. The distribution is indicated in Fig. 9. Accordingly, in making a relative evaluation of the equipment capabilities of each line, they can be broadly divided into groups (a) and (b). Furthermore, (a) can be divided into two separate characteristic groups. The equipment capabilities of each group, as well as the type of measurement methods which should be implemented, are summarized in Table 2.

The aim of the analysis in the third durability test was to examine how the variation in the result ("m" values, MTBF) from the Weibull analysis of the

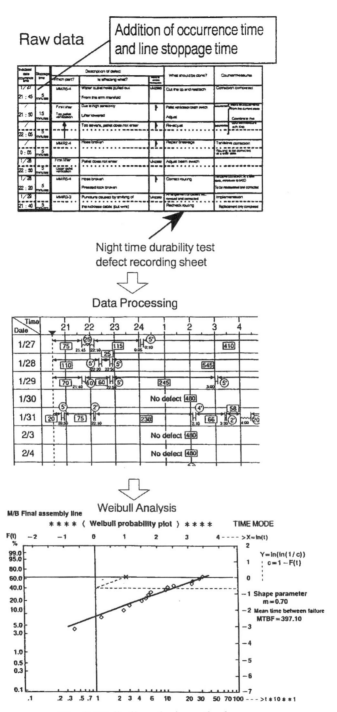

Fig. 8. Weibull analysis method

second and the third tests can affect the nature of defects and whether or not the countermeasures implemented after the second test were appropriate and effective. These results are indicated in Fig. 10 and Table 3.

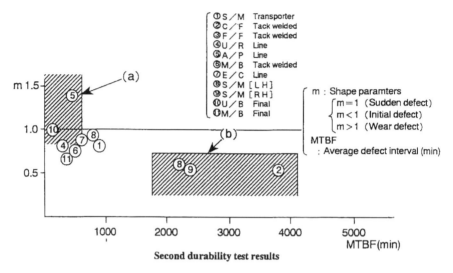

Fig. 9. "*m*" values and MTBF distribution

Table 2. Observations from analysis results: Second durability test results

Group	"m" value, MTBF	Line No.	Line Name	m	MTBF (min)	Observation from analysis results
(a) −1	m=1 (sudden defect) . MTBF is short	④	U/R	0.87	322	• Even if this had been screened out during the durability test, there is a large possibility that it may continue with a certain interval, causing the equipment to fail. • The assembly line (robots and pallets included) completion rate is low.
		⑩	U/B Final	0.99	155	
(a) −2	m>1 (wear defect) . MTBF is short	⑤	A/P	1.46	525	• Some defects can be resolved during the screening process of a durability test, while others cannot. (Most of the time, there is no resolution.) • The assembly line (robots and pallets included) completion rate is low.
(b)	m<1 (initial defect) . MTBF is short	②	C/F tack welded	0.53	3930	• There is a large possibility that the assembly line reliability can be greatly improved by screening out defects that currently occur during the durability test. • The assembly line (robots and pallets included) completion rate is high.
		③	F/F tack welded	0.60	2070	
		⑨	S/M[RH]	0.68	2061	

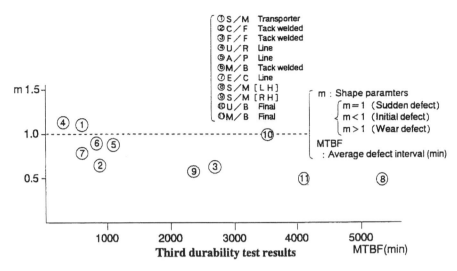

Fig. 10. "*m*" values and MTBF distribution

Table 3. Observations from analysis results: Third durability test results

Transition of MTBF	Transition of "m" values and MTBF	Line name	Observations from analysis results
Those that have improved ⑤ ⑧ ⑩ ⑪	MTBF became longer without changing m=1 (sudden defect) or m<1 (initial defect).	M/B Final	• The parity error of the I/F board of the robot made by Company B can be cited as a major defect. However, because m<1 (initial defect), the current countermeasure is thought to be effective. There is room for contemplation however, as to whether or not this countermeasure should be adopted.
		U/B Final	• The screening had been completed, and the equipment was undergoing a stable period (MTBF=280(min)).
	MTBF became longer, and m=1 (sudden defect) had changed to m>1 (wear).	A/P	• The countermeasure for the equipment defect (The countermeasure for the open circuit in the robot made by Company A) was appropriate.
	MTBF became longer, and m=1 (sudden defect) had changed to m<1 (initial defect).	S/M [LH]	• Although the reliability has improved, a new defect which is different from the one previously found may have been created. • As an applicable defect, an abnormal short circuit was being caused at a screw in the S/M robot by a mushroom-shaped chip. This must be addressed by analysis.
No changes ④⑦	m=1 (sudden defect) remains as is, and MTBF remains short.	U/R	• There is a large possibility rate later on.
Those that have worsened ①	m=1 (sudden defect) remains as is, and MTBF became shorter.	S/M transporter	• Caution ! It is very important that drastic measures are also taken by the respective manufacturers to improve their equipment's reliability.

4. 3. Detailed defect analysis

(1) Example of S/M transporter

From the second and the third durability test results, with $m=1$ (sudden defect) as is, the MTBF was shortened, requiring radical measures. The main defect was an abnormal deviation (poor stopping) of the linear transporter on the side member line. An outline of this is shown in Fig. 11.

The only countermeasure which had been taken was replacement of the instrument (encoder motor) after the second durability test, with no investigation of the real cause, even though we had to later implement a more radical measure. After the third durability test, we discovered that in the determination of the current position of the motor run, the scanning time was too long for the changes in the encoder (position detector), causing the abnormal deviation. Therefore, the radical countermeasure of reducing the scanning time to half as shown in Fig. 12 was taken.

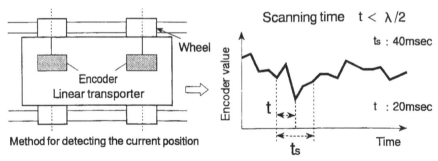

Fig. 11. Outline of a linear transporter   Fig. 12. Shortening the scanning time

(2) Example of A/P line

According to the results of the second durability test, all defects were related to the robots made by Company A. The analysis results of $m=1.46$ revealed that they were wear defects that required a radical countermeasure. Thus, since we had determined that these defects could not be remedied by screening (maintenance or replacement), it was determined to defect description to be an open circuit in the encoder.

After the second durability test, an FTA (as shown in Fig. 13) was implemented for the encoder wiring type and wire installation method used. Upon examination with a metal microscope, the female connector pin showed no damage or fretting as shown in Photo 1. However, poor pressure contact was discovered in the male connector.

We determined that this caused the resistance value to become ∞ when the wire was pulled, causing the defect to occur (abnormal deviation) (Photo 2). Later, the analysis results of the third durability test showed that the $m=1$ defect mode (sudden defect) had changed, causing the MTBF to become longer and indicating that the countermeasure was appropriate and effective.

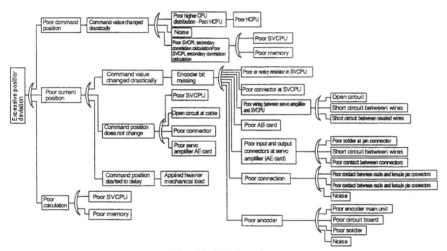

Fig. 13. FT drawing

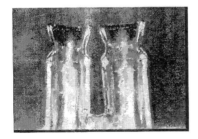

Photo 1. Female connector pin

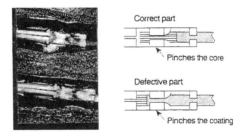

Photo 2. Male connector pin

5. Evaluation of Introduction Conditions

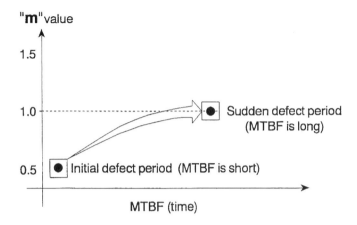

Fig. 14. Typical transition of defective condition

Table 4. Table of the rules of thumb used for analyzing the "m" values and MTBF

	Case	Changes in "m" value and MTBF	Observations of analysis results
When "m" changes	(1)	When m ≤ 1 becomes m > 1, and MTBF becomes longer	Measures taken up until now need further improvement.
	(2)	When m ≤ 1 becomes m > 1, and MTBF becomes shorter	Insufficient life for equipment. Measures taken up until now need further improvement.
	(3)	When m < 1 becomes m = 1, and MTBF becomes longer	Measures taken up until now are OK. Equipment reliability has been improved.
	(4)	When m < 1 becomes m = 1, and MTBF becomes shorter	Measures taken until now are OK. New defects have been created.
	(5)	When m ≥ 1 becomes m < 1, and MTBF becomes longer	Equipment reliability has been improved.
	(6)	When m ≥ 1 becomes m < 1, and MTBF becomes shorter	Not enough effort has been made in screening. (More has to be done before delivering them into the line.)
When "m" remains fixed	(7)	When "m" remains m > 1 and MTBF becomes longer	Only a transitory measure is taken. (A measure which addresses the real cause is not taken.)
	(8)	When "m" remains m > 1 and MTBF becomes shorter	No measure is taken.
	(9)	When "m" remains m = 1 and MTBF becomes longer	Equipment reliability has been improved.
	(10)	When "m" remains m = 1 and MTBF becomes shorter	New defects have been created.
	(11)	When "m" remains m < 1 and MTBF becomes longer	Equipment reliability has been improved.
	(12)	When "m" remains m < 1 and MTBF becomes shorter	New defects have been created.

Currently, there is no guideline which indicates how to utilize the results when applying the Weibull analysis on a system of equipment composed of numerous instruments. Therefore, we had to repeat each of the trial-and-error processes. In terms of the transition of the "m" value and MTBF, we discovered that a typical example is for $m<1$ (initial defect) with a short MTBF to ultimately turn into, $m=1$ (sudden defect) with a long MTBF as shown in Fig. 14. Although other cases are also possible, this method which monitors the "m" value and MTBF one-dimensionally and also monitors their transition, is entirely original. A summary of the rules which we are using is given in Table 4.

6. Conclusion

As a means for improving equipment reliability, we have pursued the nature of defects by conducting nighttime durability tests and applying the Weibull analysis to these defects. As a result of this analysis, we discovered that it is possible to formulate a rational and extremely effective countermeasure plan.

However, an implementation of the analysis results alone does not result in improved reliability. The importance of the final defect analysis remains unchanged, an issue that must be kept in mind. Finally, since this method is effective for all systems and instruments and can contribute greatly to production industries throughout the world, we would like to introduce it in the future to other fields.

References

[1] K. Suita, Development of Layout CAE System, Development of Process Analysis Simulator (in Japanese) *Second Production Technology Forum, Toyota Motor Corporation Production Engineering Development Committee,* 1992.

[2] Toyota Motor Corp., Twenty-fourth Toyota Reliability Professional Course Textbook (in Japanese) *Toyota Motor Corporation TQC Promotion Division,* 1992.

[3] H. Nakamura, Reliability Improvement Activity for Robots on Automobile Body Assembly Line (in Japanese) *Equipment Control Society Journal,* 1(2), 1990.

[4] H. Makabe, Introduction to Reliability Technology (in Japanese) *Japanese Standards Association,* 1984.

[5] K. Kitagawa, Introduction to Reliability Technology (in Japanese) *Corona Co.,*1979.

[6] Valter Loll, Product Certification and Standards for Testing, *Quality and Reliability Engineering International,* 10, 371-375, 1994.

[7] Valter Loll, Superimposed Renewal Processes in Repairable Systems Analysis, *Society of Reliability Engineers, Scandinavian Chapter.*

[8] Valter Loll, Handbook in Design of Reliable Equipment, *Delta Danish Electronics, Light & Acoustics,* 1993.

[9] K. Amasaka and H. Sakai, Availability and reliabilty information administration system "ARIM-BL" by methdology in "INLINE-ONLINE SQC, *International Journal of Rekiability, Quality and Safety Engineering,* 5(1), 55-63, 1998.

[10] K. Amasaka and H. Sakai, A study on TPS-QAS when utilizing Inline-Online SQC- Key to New JIT at Toyota-, *Proceedings on the Production and Operations Management Scociety, Sanfrancisco, California,* 1-8, 2002.

[11] K. Amasaka, and S. Osaki, The promotion of new statistical quality control internal education in Toyota Motor -A proposal of"Science SQC" for improving the principle of TQM-, *European Journal of Engineering Education, Research and Education in Reliability, Maintenance, Quality Control, Risk and Safety,* 24(3), 259-276, 1999.

[12] K. Amasaka, Proposal and implementation of the "Science SQC" quality control principle, *International Journal of Computer Modelling,* 38(11-13), 1125-1136, 2002.

Chapter 16: Production Process

Integration of "Inline-Online SQC"

Availability and Reliability Information Administration System "ARIM-BL"

This is a proposal for the implementation of "Science SQC" (a program based on the new scientific SQC method in Manufacturing) in order to elevate the quality of the business process. We construct the "Availability and Reliability Information Administration System - ARIM-BL" as a practical example of "Inline-Online SQC". As for the key technology of "Science SQC", it is indispensable to implement "Inline-Online SQC" by utilizing the both methods consistently. It makes the dark points in business process clear and manages to communicate between man and information. This study describes the total integrated process information system, which links together between Production Engineering division and Manufacturing Engineering division. It is representative of intelligent "Inline-Online SQC". And we have introduced the demonstrative and effective system into FBL (Flexible Body Line), which was the core of TPS (Toyota Production System).

Keywords: Reliability Information System; Availability Information System; Inline-Online SQC; Weibull Analysis.

1. Introduction

We has proposed "Science SQC", an advanced version of the SQC concept in manufacturing, as a scientific methodology to improve the quality of business process, and started research efforts to demonstrate its validity [1-6]. Our efforts in the recent years are focused on "inline SQC" in which the scientific SQC is built in the production line using the IT (Information Technology) to allow utilization of the process control information in real time, and "online SQC" in which departments concerned share and make use of the process control information.

Now, it is essential to implement "inline-online SQC" which integrates "inline SQC" and "online SQC" which are the key technologies of "science SQC" to bring unspoken information on business process to light and to comprehensively manage the "communication between man and information".

In this study, we will propose the integrated information system of intelligent "inline-online SQC" which networks the production engineering divisions and the manufacturing engineering divisions as a practical example of scientific SQC. We will also prove the effectiveness of "inline-online SQC" which was applied to the TPS (Toyota Production System)-based FBL (Flexible Body Line).

2. Effectiveness of "Inline-Online SQC"

2.1. "Inline SQC" and "Online SQC"

Practical examples of SQC include "inline SQC" which is reactively used for adaptive control of the production process [7] and "online SQC" that is shared by concerned departments and used proactively.

"Inline SQC" is exemplified by those efforts by Amasaka et al. including "mechanization of sensory inspection of axle-gear noise" [8], "mechanization of sensory inspection of absorber damping force Lissajous's waveform" [9], and "mechanization for operation intuition/knack-based corrective method for distortion of rear axle shaft" [10].

"Online SQC" includes "diagnostics for exhaust gas inspection" [11] and "method on equipment diagnosis of grinder" [12]. All of these efforts are mainly concerned with control of some specific processes.

2.2. Necessity of "Inline-Online SQC"

To achieve the purpose of the subject matter of this paper, a variety of information at the field must be collected in real time. For example, the body line involves a variety of information including the availability and defect of the "equipment", and such "production control information" as car family, vehicle model and specification. What is practiced now and then in the case of the equipment defect data, for example, is "off-line SQC" in which a member of the staff visits the site to make a record of the defect (a trouble memo), collect data and analyze the data for process improvement. However, in some cases, the repair of the defective equipment was given priority over any action needed to be taken after the hearing of defect data, and the data was not effectively used in an accurate and timely manner in the context of information management.

An essential methodology in solving these problems is "online SQC" which makes the real-time data of the "inline SQC" described in Section 2. 1 on-line on a network to allow concerned departments to process and utilize the data proactively. This will integrate the "inline SQC" and the "online SQC" as a networked system, and increase the effectiveness of the "inline-online SQC" as a scientific methodology.

3. Construction of "Availability & Reliability Information Administration System" by "Inline-Online SQC"

3.1. Network System of "Inline-Online SQC"

In order to achieve centralized control of the data concerning availability and defect of equipment by applying various "computer aids", the availability data of equipment groups in domestic and overseas plants are collected and sent via a network.

Fig. 1 shows that the availability information and the process capability information in signals are transmitted from the site information collecting PC (Personal Computer) so that each concerned department can check the information on the office monitor in real time and use the information for analysis.

3.2. Effective Introduction of "Availability & Reliability Information Administration System"

In this section, we will outline an intelligent "inline-online SQC" that was introduced and used on the FBL. At first in order to utilize this system effectively, we consider a plan how to introduce it. We propose a model to level up the information using level (IUL) gradually from "off-line SQC"(IUL-1) to "inline-online SQC"(IUL-5) as shown in Fig. 2.

Table 1 describes the information using level (IUL-1~5), SQC utilization level and each description. Under the "Availability & Reliability Information Administration System" that has been introduced, the production department analyzes the information off-line at IUL-1, and keeps track of the availability condition in-line at IUL-2. At IUL-3, the production department and the plant

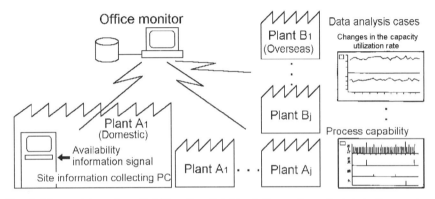

Fig. 1 Networking of availability & reliability information administration system

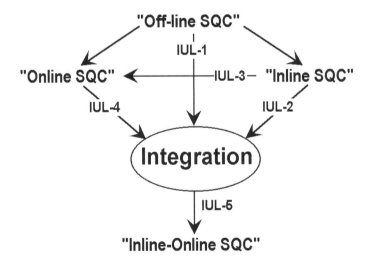

Fig. 2. Integration of "inline-online SQC"

maintenance department carry out an intelligent "inline SQC" in which the department uses the equipment defect data obtained in-line for diagnostics. At IUL-4,"online SQC" is carried out to place the information on line and make available to the production control department which will then process and utilize the information. IUL-5 aims for integration of an intelligent "inline-online SQC" in which the equipment planning department and the process design department join the network, and process and utilize the information proactively for production preparations. According to progress from IUL-1 to IUL-5 gradually, we can get a great deal of the QCD effect.

Table1. Information using level in "inline-online SQC"

Information Using Level (IUL)	SQC Utilization Level	Description of Information Provided and Utilized	QCD Effect
IUL-1	Offline SQC	A member of the production department visits the site to collect information, and the department analyzes the information. ·Equipment defect data	Poor
IUL-2	Inline SQC	Collects the availability information in-line, grasp the causalrelationship in real time and use the results for adaptive control of the process. ·Line tact, output, diagnostics, in-process control, etc.	Fair
IUL-3	Intelligent inline SQC	In addition to above, the process control condition data are provided directly to the production department, the plant maintenance management department in real time, and these departments will utilize the information proactively for process improvement. ·Equipment defect data, stratified causal information, and diagnostics information ·Defect analysis information (Weibull analysis, etc.)	Good
IUL-4	Online SQC	In addition to above, the availability information is provided in real time to the production control department, and the concerned departments in the manufacturing plant analyze the information and use the results for process improvement. ·Capacity utilization rate, MTBF, MTTR, etc.	Excellent
IUL-5	Intelligent inline-online SQC integration	In addition to above, the availability information and the equipment defect data are provided to the equipment planning department, the process design department and other upstream departments for improvement and reform of production system and process. ·Data on operation of equipment for new model	Super Excellent

3.3. Construction of "ARIM-BL" System

As a practical example of "inline-online SQC", we propose an availability & reliability information administration system "ARIM-BL" (Availability Reliability Information Monitor - Body Line) which embodies the communication between man and equipment. Fig. 3 shows the software configuration of the constructed system which was applied to FBL.

This system is consisted of (1) the body measuring system and (2) the line information processing system. The first system manages the quality inspection information including the hole location on the body and controls the accuracy of jigs used for transportation at the final process on the body production line. The second system manages the information concerning availability of the line and defect of equipment, and also carries out a reliability analysis on the defect data.

As shown above, this system makes possible a proactive centralized management of the information on the quality inspection of vehicle and the defect data on the equipment in the office, and contributes to further improvement. For example, the equipment planning department can use the system in the design review by conducting a real-time information analysis using the Weibull analysis, and the plant department can make effective use of the system in the QCD study activity for the process concerned and the PPC (Pre-Product Check) activity for the next model.

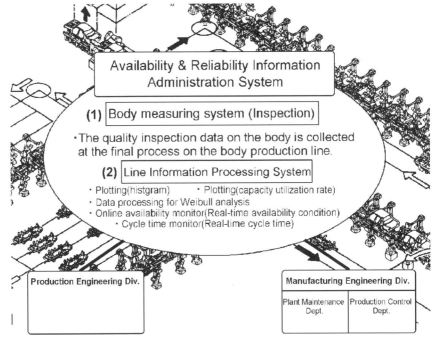

Fig.3 Example of software system for "ARIM-BL"

4. Applications

4.1. Configuration of "ARIM-BL" System

The hardware portion of the system is shown in Fig. 4. Every line in the plant is equipped with the "andon" system which manages the availability information (and especially the defect data). The information is summed up by the availability information collecting PC which is connected with the PC terminal for office managers (defect control monitor) via communication links. The PC terminal is also connected with the minicomputer for quality control which manages the vehicle inspection information (automatic body measurement data).

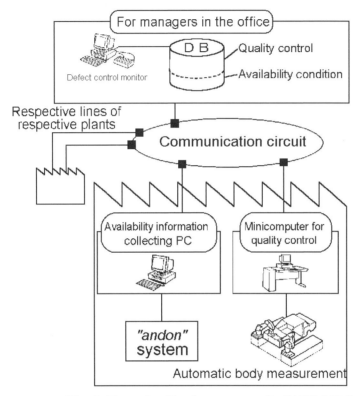

Fig. 4. Example of hardware system for "ARIM-BL"

4.2 Utilization of the System

In this section, we will describe how the availability information and the equipment defect data are collected. For example, when an equipment breaks down, the breakdown signal is collected from the "andon" control devices, and the defect distribution of each line as a function of time is outputted on the defect control monitor via communication links. This is shown in Fig. 5-(1) [13].

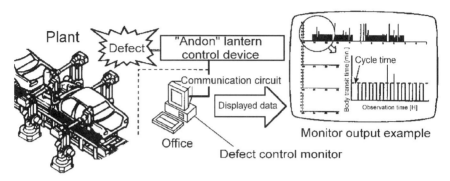

(1) How to Collect Actual Defect Data [13]

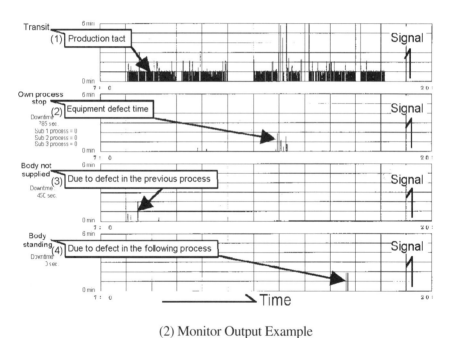

(2) Monitor Output Example

Fig. 5. Defect Data Collection Method & Example of Monitor Output

Fig. 5-(2) is an example of the output on the defect control monitor which indicates the availability of the equipment.

The monitor displays (1) production tact, (2) equipment defect time, (3) equipment defect time due to defect in the previous process and (4) equipment defect time due to defect in the following process in time series. Moreover, combination of the above information and the diagnostics information obtained by another output monitor would increase the effectiveness of the information analysis.

4.3. Effect of Introduction of the System

The following is the effect of the introduced system which utilizes the information as described in Section 4. 2. We were able to identify appropriate actions for the equipment defect data by Weibull analysis and other means. This is shown in Fig. 6. The facts mentioned above showed the conditions for analysis, and the following the results (the "m" values (shape parameters) and MTBF) in addition to the messages against defects.

Specifically, we were able to decide on appropriate actions against defects on the basis of both the "m" values and MTBF changes of each equipment in order to extend MTBF and shorten MTTR, and achieved reliability early in the stage [13]. Table 2 is a part of the law of experience which we gained by practice.

At last when we started our A1 line which was newly built to produce a new model, we effectively utilized this system.

As a result, we attained the target of line capacity utilization rate in four weeks after the line off, and met the demand for the new model.

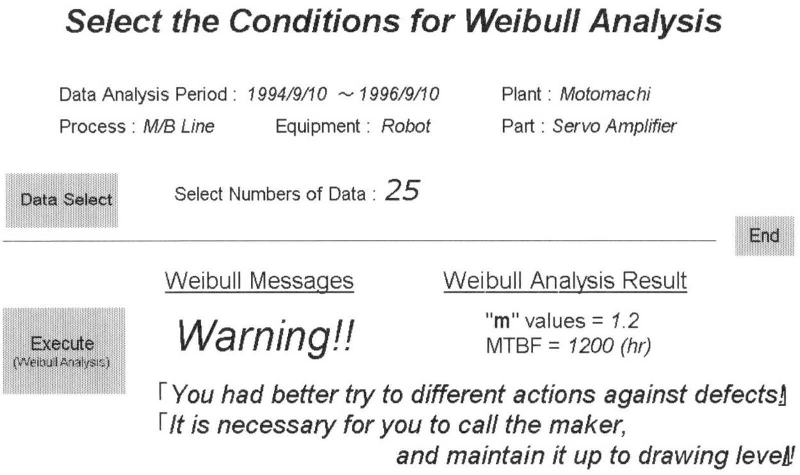

Select the Conditions for Weibull Analysis

Data Analysis Period : *1994/9/10 ~ 1996/9/10*   Plant : *Motomachi*

Process : *M/B Line*   Equipment : *Robot*   Part : *Servo Amplifier*

Data Select   Select Numbers of Data : *25*

End

Weibull Messages   Weibull Analysis Result

Execute (Weibull Analysis)   *Warning!!*   "**m**" values = *1.2*
MTBF = *1200 (hr)*

「*You had better try to different actions against defects*」
「*It is necessary for you to call the maker,*
and maintain it up to drawing level」

Fig. 6. Defect Data Select Method & Example of Monitor Output

Table 2. Experience Laws from Weibull Analysis Result

Case	"m" values changes / not change	"m" values, MTBF changes	Weibull Messages	Weibull Maintenance Messages
1	"m" values changes	"m" values changes > 1, and MTBF becomes longer.	Warning!!	You had better try to different actions against defects. It is necessary for you to call the equipment maker, and maintain it up to drawing level.
2	"m" values changes	"m" values changes > 1, and MTBF becomes shorter.	Dangerous!!	The equipment runs short of life expectancy. You had better try to different actions against defects. It is necessary for you to call the equipment maker, and maintain it up to drawing level.
3	"m" values changes	"m" values changes = 1, and MTBF becomes longer.	Best Condition!!	It is good actions against defects. Reliability of equipment improves.
4	"m" values changes	"m" values changes = 1, and MTBF becomes shorter.	Check!!	It is good actions against defects. New defects may occur.
5	"m" values changes	"m" values changes < 1, and MTBF becomes longer.	Good!!	Reliability of equipment improves. You had better keep on present actions.
6	"m" values changes	"m" values changes < 1, and MTBF becomes shorter.	Do your best!!	You must make an effort at screening!! You have to do it before you introduce equipment into the line.
. . .				

5. Conclusion

By constructing the "Availability & Reliability Information Administration System - ARIM-BL" and applying the system to FBL, we have improved the QCD effect, confirmed the effectiveness of inline-online SQC and made various achievements as shown above.

References

[1] Amasaka, K (1998): Application of Classification and Related Methods to the SQC Renaissance in Toyota Motor, Hayashi,C et al.(Eds.), *Data Science, Classification and Related Methods,* 684-695, *Springer.*

[2] K. Amasaka(1999): A study on"Science SQC"by Utilizing"Management SQC, -Demonstrative Study on a New SQC Concept and Procedure in the Manufacturing Industry-, *Journal of Production Economics,* 60-61, 591-598.

[3] Amasaka, K and Osaki, S (1999): The Promotion of New Statistical Quality Control Internal Education in Toyota Motor -A Proposal of"Science SQC" for

Improving the Principle of TQM-, *European Journal of Engineering Education (EJEE), Research and Education in Reliability, Maintenance, Quality Control, Risk and Safety,* 24(3), 259-276.

[4] Amasaka, K(2000): A Demonstrative Study of a New SQC Concept and Procedure in the Manufacturing Industry, -Establishment of a New Technical Method for Conducting Scientific SQC-, *An International Journal of Mathematical & Computer modeling,* 31(10-12), 1-10.

[5] Amasaka, K(2002):Proposal and Implementation of the"Science SQC"Quality Control Principle, *International Journal of Computer Modeling.* (decided to be published)

[6] Amasaka, K and Osaki, S (2002): A Reliability of Oil Seal for Transaxle- A Science SQC Approach in Toyota-, *Case Studies in Reliability and Maintenance by Wallace R.Blischke and D. N. P. Murthy, to be published by John Wiley & Sons, Inc.,* 571-581.

[7] M. Matsubara et al.(1980):Flexible Measuring Unit, (in Japanese) *TOYOTA Technology Conference,* 29(1), 45-49.

[8] K. Amasaka et al.(1972):Mechanization of Sensory Inspection of Axle-gear Noise, (in Japanese) *Proceedings on the JUSE(Union of Japanese Scientists and Engineers), The 2nd Sensory Inspection Symposium,* 5-12.

[9] K. Amasaka and H. Saito(1983):Mechanization of Sensory Inspection of Absorber Damping Force Lissajous's Waveform, *The Chubu Association for Quality Control of Japan, QC Technical Conference,* 48-54.

[10] K. Amasaka(1983):Mechanization for Operation Depended on Intuition/knack-based Corrective Method for Distortion of Rear Axle Shaft, *JSQC (Journal of the Japanese Society for Quality Control), 26th Technical Conference,* 5-10.

[11] K. Amasaka and T. Kosugi(1991):Application and Effects of Multivariate Analysis in TOYOTA, (in Japanese) *The Behavior Metric Society of Japan, 19th Annual Conference,* 178-183.

[12] K. Amasaka et al.(1992):Method on Equipment Diagnosis of Grinder, *JSQC(Journal of the Japanese Society for Quality Control), 42th Technical Conference,* 37-40.

[13] K. Amasaka and H. Sakai(1996):Improving the Reliability of the Body Assembly Line Equipment, *International Journal of Reliability, Quality and Safety Engineering,* 3 (1), 11-24.

Chapter 17: Process Control

TPS Manufacturing Fundamentals:

Applying Intelligence Control Charts by "TPS-QAS" with "Inline-Offline SQC"

The matter of course that manufacturing should follow the instruction specified by the drawing is one of the fundamentals for manufacturing quality control. However, to the author's knowledge, due to the advancement of mechanization, it seems that control charts are not necessarily utilized to the full extent by a number of enterprises in their process control. In recognition of the importance of a scientific process control as the fundamental of quality control, the author proposes "TPS-QAS". TPS-QAS is a new scientific process control method with so-called intelligence control charts that utilizes the "Inline-Online SQC" that exploits IT and SQC through the application of the "Science SQC". This paper demonstrates the effectiveness of the proposed process control by quoting the demonstrative cases at Toyota Motor Corporation.

Keywords; Intelligence Control Charts, TPS-QAS, Inline-Online SQC, Science SQC, Toyota's TQM activities.

1. Introduction

To satisfy customers in the world with quality products, Toyota continues to build in quality on the line as a manufacturer. This is one of the fundamentals of Toyota's TQM activities[1-2]. To the author's knowledge, due to the advancement of mechanization, it seems that control charts are not necessarily utilized to the full extent by a number of enterprises in their process control [3].

In recognition of the importance of a scientific process control as the fundamental of quality control, the author proposes "TPS-QAS (Toyota

Production System-Quality Assurance by Utilizing Science SQC)". TPS-QAS is a new scientific process control method with so-called intelligence control charts that utilizes the "Inline-Online SQC" that exploits IT and SQC through the application of the "Science SQC"[4-6].

This paper demonstrates the effectiveness of the proposed process control by quoting the demonstrative cases at Toyota Motor Corporation [7].

2. Necessity of Scientific Process Control

2.1. Background [3]

(Case 1)

In the past investigation, there were several cases in which the cause of a market claim was found to be the insufficient capability of the machining process. In addition, in those cases, it was found that the machining rate of the production line had been raised to meet high production load.

Reviewing the operation drawing (machining standard) issued by the pertinent production engineering division revealed that the manufacturing division changed the manufacturing conditions including the "machining rate"-an important factor in retaining the process capability-without updating the operation drawing.

Such an example implies the importance of scientific process control. In other word, the production line should accompanied by reasonable quality control-returning to the fundamentals of quality-first CS activities and ensuring to build in quality in the line.

(Case 2)

Conventionally, control charts are drawn manually and without sufficient considerations. Thanks to advancement of automatization, one may say, "No problem. We have a fool-proof mechanism," or "Our automated inspection system is sufficient for building in quality in the line" and so forth.

However, we must remember that the foolproof mechanism is only a method for detecting abnormality and not a method for dispersion-proof that ensures process capability. Similarly, the automated inspection system is only a method for consecutive adjustment and not necessarily designed as a process control method for stable production.

In summary, we have to reconsider our process control, which has become too dependent on the equipment due to the introduction of automatic inspection system or automation of the equipment. Our system has to be linked to the organization-wide activities to maintain operators' motivation to build in quality in the line.

2.2. Importance of Control Chart [7]

"Manufacturing" is a "battle against dispersion" and the real cause of dispersion, if any, has to be detected and controlled. Quality problems of manufacture (dispersion) appear as the result of combined dispersions in the process of work (cause). Consequently, it is important to utilize the control chart as one of the scientific tools to visualize the quality of manufacture and realize the importance of quality of manufacture (workmanship).

For the success of an organization-wide activity, information Technology (IT) and SQC have to be fully utilized to share the process quality control information among the design, production engineering and manufacturing divisions. If design quality or manufacturing quality should fail to achieve quality requirements of the customers, 5M-E (Man, Machine, Material, Method, Measuring and Environment) and other elements have to be investigated, recurrence be prevented, work process be improved to prevent bias and/or dispersion to target quality and the process capability (Cp) and machine capability (Cm) be retained.

3. Proposal for "TPS-QAS" Using "Inline-Online SQC" [7]

3.1. Necessity of "Inline-Online SQC"

To build in stable quality in the line, it is not enough for the line operator and/or staff to carry on their job by solely depending on "Offline SQC" such as desktop drawing of a control chart or data analysis for improvement. They have to also gather various workplace information promptly for analysis.

For example, "quality control information" is congested with information on parts (suppliers), production equipment operation and defects, whereas "production control information" also includes wide-ranging information such as the vehicle type, model and specifications. Under the general circumstances, pre-production check prior to the next model launch is often depending on afterwards manual data collection and analysis-collection of data from trouble reports and on-site interview by technical staff visit. With this approach, it is difficult to gather sufficient amount of accurate data in a timely manner.

In this sense, through utilization of IT, the ideal form of the future quality control should enable processing of real-time data and promote Inline SQC for feedback and/or feed-forward, and Online SQC for disclosing and sharing information among related divisions. In other words, the author believes that it is necessary to make the information accessible by the managers and staff of the line and the design, production engineering and manufacturing divisions. In addition, Intelligent Inline-Online SQC that enables all relevant people to work on problem solving through cause analysis of the shared data as organization-wide.

For instance, this means that information originating from the line is gathered with personal computers and transferred to other related divisions. With this system, anyone can carry on process improvement promptly as may be needed by reading the quality control information of the process anytime, anywhere. Such improvement will eventually improve the process and machine capability (Cp & Cm) as the organization-wide activities.

The author proposes, "TPS-QAS", a scientific process control that utilizes IT and SQC to facilitate timely browsing of manufacturing data on the production line.

3.2. Proposal of "TPS-QAS"

Fig. 1 is a conceptual diagram of "TPS-QAS", which can realize the Inline-Online SQC described earlier. As Fig. 1 indicates, Offline SQC saves the tedious work of drawing a control chart with user-friendly SQC software, "TPOS". Time saving allows more time for decision making and diagnosis by the line managers and engineering staff, helping to improve the overall performance of the workplace. In addition, Inline SQC enables real time grasping of quality control status in the process and accurate diagnosis that leads to prompt actions.

On the other hand, the design, production engineering and manufacturing divisions operate the Online SQC (as proposed in Fig. 1) online. Table 1 expressly outlines the systematic development of the manufacturing quality assurance level (QA level) for realizing "TPS.S"[4] (TPS by utilizing Science SQC) in the case of control chart application.

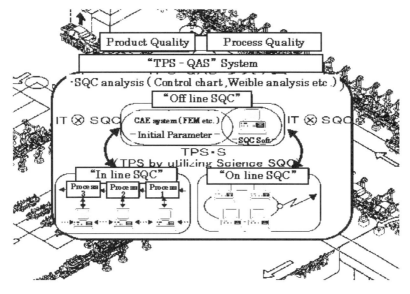

Fig. 1. Shematic Drawing of "TPS-QAS" [7]

Table 1. Systematic utilization level of quality control information [7]

QA level	Sample development of a production control chart
QA level 1	Production control chart (hand written)
QA level 2	Production control chart (using personal computer) ━━▶ Improved Cm, Cp (1)
QA level 3	Joint utilization of control chart by production and maintenance divisions ━━▶ Improved Cm, Cp (2) Information feedback to maintenance
QA level 4	Use of QA level 3 information by production engineering divisions ━▶ Improved equipment reliability and maintain ability Information feedback to process design, etc
QA level 5	Use of QA level 3 information by engineering design divisions ━▶ Information feedback to design drawings, etc
QA level 6	Use of QA level 3 information by the main suppliers, etc ━▶ Information feedback to purchased parts for improved Cm, Cp (3)
QA level 7	Use of QA level 3 information by the main sales and service shops ━▶ Information feedback to the main shops for building ties with customers
QA level 8	Introduce QA level 3 information to overseas plants.

4. Cases of Application [7] [8]

4.1. Application to the model line (Mountain climbing figure)

The mountain climbing diagram as shown in Fig. 2 is a time series indication of the system introduction (mountain climbing) as proposed in Fig. 1.Starting at the end of 1997, the author selected a model line, developed the environment and software and proceeded to introducing the system (hardware). At present, the author is promoting the utilization of the Inline-Online SQC through introduction to all divisions of the company.

A model line for the trial of this system should have both of the following features:

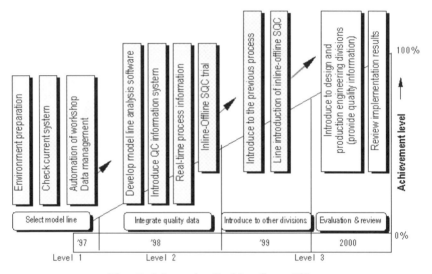

Fig . 2. Mountain climbing figure [8]

(1) Manual line by operators
(2) Automated line

A manual line by operators (1) employs either manual data input by a man or automatic data input with signals. Considering this point, the Inline-Online SQC system was designed to have the general-purpose portion (main system) and dedicated portion (subsystem) to make it applicable to both Cases (1) and (2). The example described later will mainly apply to Case (2).

4.2. Quality Control Information System on Model Line

An automated assembly line-applicable to the above (2)-was selected as the model line. The line assembles a differential unit that transmits driving power from the engine to the tires. Fig. 3 shows the hardware system of "TPS-QAS". The general-purpose main system can be applied to Case (1), manual line by operators. On the other hand, the subsystem is applied to Case (2), the automated line. In the automated line, works on pallets proceed through the assembly processes. At each process, running torque and other process control information is sequentially written on the ID tag (①) attached to the palette. At the final

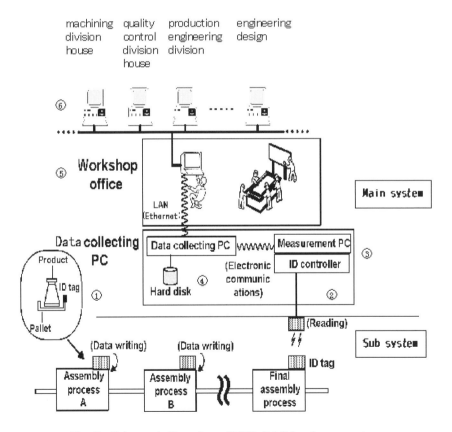

Fig. 3. Schematic Drawing of TPS-QAS hardware system

assembly process, the ID tag information is processed to ID controller (②) and measurement PC (③) and stored in the data collecting PC (④).

The "TPS-QAS" connects the data collecting PC to the terminals of the manufacturing workshop office (⑤), machining division house, quality control division house, production engineering and engineering design divisions(⑥) in a network. It is the inline system that enables real-time information disclosure and utilization. It further discloses information to related divisions through the online system.

4. 3. Utilization example of intelligence control chart with "TPS-QAS" [7]

In order to minimize time lag for disclosure and utilization of process control information, in the TPS-QAS, in-line data stored in the highly advanced database are obtained through continual sampling. Through automatic creation of intelligence control charts, process control status and abnormality are monitored real time, making it possible for the process to take a prompt remedy action.

From the model line, about 500 intelligence control charts (10 control items/unit x 5 units/ line x 10 lines/plant) were created for QA level 4 of Table 1. These intelligence control charts are accessible from related divisions for monitoring process control state, process capability, and timely problem detection and solution.

The system features improved operability for selecting quality characteristics and extended functionality. Necessary control characteristics can be specified according to the stratified structure. The user selects from or enters to (A) Select Item, (B) Select Detailed Item, and (C) Extract conditions as indicated Fig. 4(a). The features of the intelligence control chart software are indicated in Fig. 4(b).

As the support functions for diagnosing a process and taking a remedy action according to the findings obtained from the intelligence control chart, (1) Scroll function and (2) Raw data conversion function are newly added. These functions help improve the process by converting the sample data into an appropriate form which is more suitable for analysis. Regarding (3) Stratified cause analysis, data is sort out based on stratification to trace the causal relation of the factors. In addition, past know-how is utilized by developing (4) Improvement history database, and a warning is generated automatically on abnormal process by using IT to extract processes with and without abnormality at random. (5) Abnormality diagnosis function makes efficient operation possible. Furthermore, (6) Data link with other application software utilizes TPOS analysis software to enable real-time cause analysis.

Through the actual application of the intelligence control chart software, the author received user feedbacks such as (a) "Scroll function is useful for reviewing the past data", (b) "Automatic warning is very effective as it only warns abnormal control characteristics", (c) "Raw data conversion and stratified cause analysis are helpful for cause analysis."

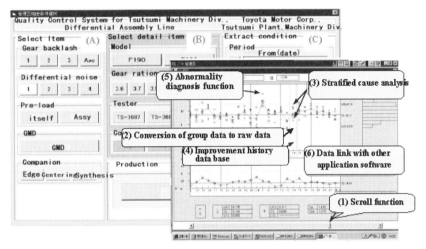

Fig. 4. Outline of Intelligence Control Chart Utilizing Software System [7]

4.4. Process improvement example of intelligence control chart

Fig. 5 is the X-R control chart of secondary vibration for differential connection, which is an important quality characteristic for differential noise control. Differential noise is caused by the vibration at the connection between the differential unit drive pinion and ring gear due to insufficient machining accuracy at the connection. Differential noise is heard as unpleasant noise at 80 km/hr. As this noise is also affected by the installation quality of differential unit sub-components, the system was utilized thoroughly for the all differential unit assembly processes.

Process survey and cause analysis revealed insufficient process capability of the secondary vibration for differential connection. A number of cause analyses found low process capability for gear accuracy as indicated in Fig.5-(a), which was caused by insufficient maintenance for the drive pinion machining equipment. After identification of optimal machining conditions and adjustment, the process capability was improved as indicated in Fig. 5-(b).

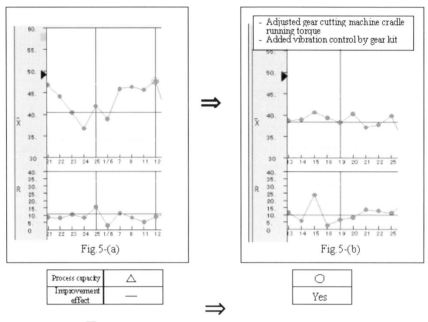

Fig. 5 $\overline{X}$-R control chart of secondary vibration at differential connection [8]

4.5. Organizational application of the system

In recent years, Toyota Motor Corporation is promoting the introduction of the system to all Toyota plants in Japan. The system is also introduced to other Toyota group companies and overseas plants. In the separate study, the author proposed and verified the effectiveness of a equipment failure information diagnosis system, "ARIM-BL (Availability and Reliability Information Monitor System Body Line) [9]. Further study proved that linking the TPS-QAS with the ARIM-BL enhances the intelligence control chart reliability, signifying the TPS-QAS effectiveness.

5. Conclusion

The matter of course that manufacturing should follow the instruction specified by the drawing is one of the fundamentals for manufacturing quality control. To fulfill the fundamental, it is clear that utilization of the intelligence control charts, as a scientific process control tool is essential for process control. Returning to the fundamentals of manufacturing and promoting the intelligence control chart utilization, the author will further continue introduction of the Inline-Offline SQC to advance scientific process control suitable for the future manufacturing.

References

[1] K.Amasaka, (2002), "Science TQM, A New Principle for Quality Management-A Demonstrative Study through New TQM Activities at Toyota-", *2nd Euro-Japanese Workshop on Stochastic Risk Modeling for France; Insurance, Production and Reliability, Chamonix, France,* pp.6-14.

[2] K.Amasaka, (1989), "TQC at Toyota, Actual State of Quality Control Activities in Japan", *Union of Japanese Scientists and Engineers, The 19th Quality Control Study Team of Europe,* 39, pp.107-112.

[3] Edited by K.Amasaka, (2003), "Manufacturing Fundamental: The Application of Intelligence Control Charts-Digital Engineering for Superior Quality Control-", (in Japanese) *Japanese Standards Association.*

[4] K.Amasaka, (1999), "A study on 'Science SQC' by Utilizing 'Management SQC'- A Demonstrative Study on a New SQC Concept and Procedure in the Manufacturing Industry-", *Journal of Production Economics,* Vol.60-61, pp.591-598.

[5] K.Amasaka, (2000), "A Demonstrative Study of a New SQC Concept and Procedure in the Manufacturing Industry-Establishment of a New Technical Method for Conducting Scientific SQC-", *An International Journal of Mathematical & Computer modeling,* Vol.31, No.10-12, pp.1-10.

[6] K.Amasaka, (2003), "Proposal and Implementation of the 'Science SQC' Quality Control Principle", *International Journal of Mathematical and Computer Modeling, Vol. 38, No. 11-13, pp. 1125-1136.*

[7] K.Amasaka and H. Sakai, (2002), "A Study on TPS-QAS When Utilizing Inline-Online SQC-Key to New JIT at Toyota-", *Proceedings of the Production and Operations Management Society, POMS2002 JIT Manufacturing/ Lean*

Production Sessions, San Francisco, California, pp.1-8.

[8] Edited by K.Amasaka,(2000), "Science SQC: The Quality Reform of the Business Process", (in Japanese) *Japanese Standards Association.*

[9] K.Amasaka and H.Sakai, (1998), "Availability and Reliability Information Administration System, 'ARIM-BL' by Methodology in 'Inline-Online SQC', *International Journal of Reliability, Quality and Safety Engineering,* Vol.5, No.1. pp.55-63.

5. Conclusion

5. Conclusion

Looking back at the change in atmosphere that took place last year in manufacturing, the need to improve the technical ability of manufacturers and the increase in effort by product developers is evident. The important thing is to once again realize the importance of Statistical Quality Control (SQC) in proactive technology, instead of dealing with reactive technology.

Not only should SQC be applied to general statistical analysis, it should also be used from a scientific perspective. Furthermore, the intricate cause-and-effect relationship to analyze the gap between reality and theory should be examined, new truths and knowledge uncovered, and answers found through a "basic method", a method applicable universally, instead of settling for an "individual method", to find special or partial solutions.

To create a "basic method" proactive technological solution, specialized SQC technology must be avoided by establishing systematic research for organizational practices, creating a propulsion cycle that maximizes technology, and applying these to the business and manufacturing process. The author has, therefore, proposed a quality control principle, "Science SQC" [1] as a demonstrative-scientific methodology and discussed the effectiveness of this method, which improves the systematic development of the principle of TQM.

"Science SQC" is a new system of SQC application in which four core principles and linked. The first principle, "Scientific SQC", is a scientific approach, while the second principle, "SQC Technical Methods", is a methodology for solving problems. The integrated SQC network system "TTIS: Total SQC Technical Intelligence System" is the third principle. It transforms technologies into assets for intellectual networking to ensure scientific evolution. "Management SQC" is the fourth principle. It is used to determine technical gaps between theories and facts, as organizational problems between departments change implicit knowledge of inherent business processes into explicit knowledge for general solutions.

Toyota Motor Corporation developed "Science SQC", as proposed by the author [2-3], to plan strengthening of the technological quality strategy with the aim of becoming the leader in the business administration in the 21st century. Toyota has demonstrated that this methodology can improve the quality of engineers' work at every stage of the business process, and contribute to creating products of excellent quality. Furthermore, to make "Science SQC" continuous with future developments, it is important to keep the SQC promotion cycle, consisting of implementation, practical effort, education and growth of human resources, rotating in order to achieve steady development of company-wide SQC promotion activities.

In the future, "Science SQC" will be positioned as a new quality control and applied to solving various practical problems. With the accumulation of demonstrative studies, the author strongly desire that next generation TQM, "Science TQM" [4-5], or so-called "TQM-S" (TQM utilizing "Science SQC") [6], will be established at Toyota to improve the principle of TQM.

References

[1] K. Amasaka, (2003), Proposal and Implementation of the "Science SQC" Quality Control Principle, *International Journal of Mathematical and Computer Modelling,* Vol. 38, No. 11-13, pp. 1125-1136.

[2] K. Amasaka, (1999), A study on "Science SQC" by Utilizing "Management SQC": A Demonstrative Study on a New SQC Concept and Procedure in the Manufacturing Industry, *International Journal of Production Economics,* Vol. 60-61, pp. 591-598.

[3] K. Amasaka and S. Osaki, (1999), The Promotion of New Statistical Quality Control Internal Education in Toyota Motor: A Proposal of "Science SQC" for Improving the Principle of TQM, *The European Journal of Engineering Education,* Vol. 24, No. 3, pp. 259-276.

[4] K. Amasaka, (2002), Science TQM, A New Principle for Quality Management: A Demonstrative Study through new TQM Activities at Toyota, *The 2nd Euro-Japanese Workshop on Stochastic, Chamonix, France,* pp. 6-14.

[5] K. Amasaka, (2003), Development of "Science TQM", A New Principle of Quality Management: Effectiveness of Strategic Stratified Task Team at Toyota, *Proceedings of the 17th International Conference on Production Research, Blacksburg, Virginia,* pp.1-10, *(To be published in the International Journal of Production Research, 2003)*

[6] K. Amasaka, (2000), "TQM-S", A New Principle for TQM Activities: A New Demonstrative Study on "Science SQC", *Proceedings of the International Conference on Production Research, Bangkok, Thailand,* pp. 1-6.

Index